ADVANCED ELECTROCULTURE: HARNESSING EARTH BATTERIES AND ELECTROCULTURE

DAVID HOLMAN

TABLE OF CONTENTS

DAVID HOLMAN

DAVID HOLMAN

DAVID HOLMAN

DAVID HOLMAN

INTRODUCTION

WELCOME TO THE WORLD OF
ADVANCED ELECTROCULTURE,
WHERE SCIENCE, SUSTAINABLE
AGRICULTURE, AND INNOVATION
CONVERGE TO REVOLUTIONIZE
THE WAY WE GROW OUR FOOD.

DAVID HOLMAN

IN THE FOLLOWING PAGES, WE WILL EMBARK ON A JOURNEY THAT DELVES DEEP INTO THE INTRICACIES OF ELECTROCULTURE, EXPLORING ITS RICH HISTORY, ITS PROFOUND CONNECTION TO PLANT GROWTH, AND THE GROUNDBREAKING INTEGRATION OF EARTH BATTERIES AND CAPACITANCE SYSTEMS TO POWER AND ENHANCE AGRICULTURAL PRACTICES.

THE PROMISE OF ADVANCED ELECTROCULTURE

IN AN ERA DEFINED BY ENVIRONMENTAL CHALLENGES AND THE PRESSING NEED FOR SUSTAINABLE SOLUTIONS, ADVANCED ELECTROCULTURE STANDS AS A BEACON OF HOPE.

IT OFFERS A NEW PARADIGM FOR FARMING THAT NOT ONLY MAXIMIZES CROP YIELDS BUT

ALSO MINIMIZES THE ENVIRONMENTAL IMPACT OF AGRICULTURE. AT ITS CORE, ELECTROCULTURE HARNESSES THE POWER OF ELECTRICITY TO STIMULATE PLANT GROWTH, BOOST NUTRIENT UPTAKE, AND COMBAT PESTS AND DISEASES.

IT PROMISES HIGHER AGRICULTURAL PRODUCTIVITY, REDUCED CHEMICAL USAGE, AND A MORE ENVIRONMENTALLY FRIENDLY APPROACH TO FOOD PRODUCTION.

AS WE DELVE DEEPER INTO THE WORLD OF ADVANCED ELECTROCULTURE, WE WILL UNVEIL THE SECRETS BEHIND THIS REMARKABLE SYNERGY BETWEEN ELECTRICITY AND AGRICULTURE. YOU WILL DISCOVER HOW PIONEERS IN THIS FIELD HAVE UNLOCKED THE POTENTIAL OF ELECTROCULTURE TO CULTIVATE HEALTHIER CROPS, CONSERVE

RESOURCES, AND BUILD A MORE SUSTAINABLE FUTURE FOR OUR PLANET.

SUSTAINABLE ENERGY AND AGRICULTURE

AT THE HEART OF ADVANCED ELECTROCULTURE LIES THE CONCEPT OF SUSTAINABILITY. AS OUR GLOBAL POPULATION CONTINUES TO GROW, THE DEMAND FOR FOOD PRODUCTION INCREASES EXPONENTIALLY. HOWEVER, TRADITIONAL FARMING PRACTICES OFTEN COME AT A HIGH COST TO THE ENVIRONMENT. THEY CONTRIBUTE TO SOIL DEGRADATION, WATER POLLUTION, AND GREENHOUSE GAS EMISSIONS.

IN THE FACE OF THESE CHALLENGES, WE FIND OURSELVES AT A CRITICAL JUNCTURE WHERE WE MUST REIMAGINE

AGRICULTURE FOR A SUSTAINABLE FUTURE.

THIS BOOK EXPLORES THE INTERSECTION OF ENERGY AND AGRICULTURE, OFFERING A VISION WHERE THE TWO WORLDS HARMONIZE TO CREATE A SYMBIOTIC RELATIONSHIP. EARTH BATTERIES, A RELATIVELY UNTAPPED ENERGY SOURCE, BECOME AN INTEGRAL PART OF THIS VISION.

THEY PROVIDE A RENEWABLE AND ENVIRONMENTALLY FRIENDLY ENERGY SUPPLY TO POWER ELECTROCULTURE SYSTEMS, REDUCING THE RELIANCE ON NON-RENEWABLE ENERGY SOURCES AND MINIMIZING THE CARBON FOOTPRINT OF FARMING.

THE CHAPTERS THAT FOLLOW WILL GUIDE YOU THROUGH THE INTRICACIES OF THIS

DAVID HOLMAN

SUSTAINABLE ENERGY-AGRICULTURE NEXUS.

WE WILL WALK THROUGH THE HISTORICAL CONTEXT OF ELECTROCULTURE, EXAMINE THE FASCINATING ELECTRIC CONNECTION TO PLANT GROWTH, AND DELVE INTO THE PRINCIPLES OF EARTH BATTERIES AND CAPACITANCE. TOGETHER, WE WILL UNLOCK THE POTENTIAL OF THESE TECHNOLOGIES AND EXPLORE THEIR TRANSFORMATIVE IMPACT ON MODERN AGRICULTURE.

NAVIGATING THE JOURNEY AHEAD

OUR EXPLORATION OF ADVANCED ELECTROCULTURE IS STRUCTURED METICULOUSLY TO PROVIDE YOU WITH A COMPREHENSIVE

UNDERSTANDING OF THIS EMERGING FIELD.

EACH CHAPTER IS DESIGNED TO BUILD UPON THE KNOWLEDGE GAINED IN THE PREVIOUS ONE, CREATING A SEAMLESS LEARNING EXPERIENCE.

WHETHER YOU ARE A SEASONED FARMER, AN ASPIRING AGRICULTURALIST, OR SIMPLY CURIOUS ABOUT THE POSSIBILITIES OF ELECTROCULTURE, THIS BOOK IS TAILORED TO MEET YOUR NEEDS.

IN CHAPTER 1, WE WILL VENTURE INTO THE HISTORICAL CONTEXT OF ELECTROCULTURE, TRACING ITS ROOTS FROM ANCIENT PRACTICES TO MODERN-DAY APPLICATIONS. WE WILL UNCOVER HOW EARLY FARMERS STUMBLED UPON THE BENEFICIAL EFFECTS OF ELECTRICITY ON PLANT GROWTH AND SET THE STAGE FOR THE

SCIENTIFIC DISCOVERIES THAT
FOLLOWED.

CHAPTER 2 WILL IMMERSE YOU IN
THE WORLD OF EARTH BATTERIES.
WE WILL DEFINE WHAT EARTH
BATTERIES ARE, EXPLORE THEIR
PRINCIPLES, AND DIVE INTO THE
FASCINATING REALM OF
HARNESSING EARTH BATTERIES
FOR FARMING.

YOU WILL BE INTRODUCED TO
REAL-WORLD CASE STUDIES
WHERE EARTH BATTERIES HAVE
YIELDED REMARKABLE RESULTS
IN AGRICULTURAL SETTINGS.

THE EXPLORATION CONTINUES IN

CHAPTER 3, WHERE WE UNRAVEL
THE CONCEPT OF CAPACITANCE IN
AGRICULTURE.

WE WILL UNDERSTAND THE ROLE
OF CAPACITANCE IN ELECTRICAL
SYSTEMS AND HOW IT CAN BE

EMPLOYED TO ENHANCE ELECTROCULTURE SYSTEMS.

YOU WILL GAIN INSIGHTS INTO THE VARIOUS CAPACITANCE-BASED TECHNOLOGIES THAT ARE TRANSFORMING THE AGRICULTURAL LANDSCAPE.

CHAPTER 4 IS A PIVOTAL POINT IN OUR JOURNEY, WHERE WE EXAMINE THE POTENTIAL SYNERGY BETWEEN EARTH BATTERIES AND CAPACITANCE.

WE WILL UNCOVER THE ADVANTAGES OF INTEGRATING THESE TECHNOLOGIES, CREATING SYSTEMS THAT ARE NOT ONLY SUSTAINABLE BUT ALSO HIGHLY EFFICIENT IN PROMOTING PLANT GROWTH AND CROP HEALTH.

IN CHAPTER 5, WE WILL EMBARK ON A PRACTICAL JOURNEY, OFFERING GUIDANCE ON

DESIGNING AND BUILDING ELECTROCULTURE SYSTEMS.

YOU WILL RECEIVE STEP-BY-STEP INSTRUCTIONS, TO HELP YOU CONSTRUCT YOUR OWN ELECTROCULTURE SYSTEM, CUSTOMIZED TO YOUR SPECIFIC NEEDS.

THE JOURNEY INTO THE WORLD OF ADVANCED ELECTROCULTURE CONTINUES WITH....

CHAPTER 6, WHERE WE EXPLORE THE INTRICACIES OF BUILDING EARTH BATTERIES.

WE WILL DELVE INTO THE MATERIALS REQUIRED, DESIGN CONSIDERATIONS, AND SAFETY MEASURES NECESSARY TO CONSTRUCT EARTH BATTERIES EFFECTIVELY.

CHAPTER 7 TAKES US FURTHER INTO THE PRACTICAL REALM, GUIDING YOU ON HOW TO CONNECT EARTH BATTERIES TO

ELECTROCULTURE SYSTEMS SEAMLESSLY.

WE WILL DISCUSS WIRING CONFIGURATIONS, OPTIMIZATION TECHNIQUES, AND PROVIDE PRACTICAL EXAMPLES TO ENSURE A SMOOTH INTEGRATION PROCESS.

MONITORING AND MAINTENANCE ARE CRITICAL ASPECTS OF SUSTAINING ELECTROCULTURE SYSTEMS, AND....

CHAPTER 8 IS DEDICATED TO THESE TOPICS. YOU WILL LEARN HOW TO MONITOR SYSTEM PERFORMANCE AND IMPLEMENT ROUTINE MAINTENANCE TO ENSURE YOUR SYSTEM OPERATES AT PEAK EFFICIENCY.

IN CHAPTER 9, WE WILL STEP INTO THE REAL WORLD, EXPLORING CASE STUDIES OF FARMERS AND RESEARCHERS WHO HAVE SUCCESSFULLY IMPLEMENTED

ADVANCED ELECTROCULTURE SYSTEMS.

THESE STORIES WILL PROVIDE YOU WITH TANGIBLE EXAMPLES OF THE OUTCOMES AND BENEFITS ACHIEVABLE THROUGH ELECTROCULTURE.

CHAPTER 10 TACKLES THE CHALLENGES YOU MAY ENCOUNTER ON YOUR ELECTROCULTURE JOURNEY AND OFFERS INNOVATIVE SOLUTIONS AND WORKAROUNDS TO OVERCOME THEM.

WE BELIEVE THAT ADDRESSING CHALLENGES HEAD-ON IS ESSENTIAL FOR THE CONTINUED ADVANCEMENT OF THIS FIELD.

AS WE LOOK TO

CHAPTER 11, WE WILL EXAMINE EMERGING TECHNOLOGIES AND

TRENDS IN ELECTROCULTURE AND SUSTAINABLE AGRICULTURE.

THE POSSIBILITIES ARE BOUNDLESS, AND WE WILL SPECULATE ON THE EXCITING DEVELOPMENTS THAT LIE AHEAD.

FINALLY, IN

CHAPTER 12, WE WILL BRING OUR JOURNEY TO A CLOSE. YOU WILL FIND A SUMMARY OF KEY TAKEAWAYS, ALONG WITH WORDS OF ENCOURAGEMENT TO INSPIRE YOU TO IMPLEMENT ADVANCED ELECTROCULTURE IN YOUR FARMING PRACTICES.

TOGETHER, WE WILL CHART A PATH TOWARD A MORE SUSTAINABLE AND PROSPEROUS AGRICULTURAL FUTURE.

EXPLORING THE POTENTIAL

IN THE PAGES THAT FOLLOW, YOU WILL FIND A WEALTH OF

KNOWLEDGE, PRACTICAL INSIGHTS, AND INSPIRATION. ADVANCED ELECTROCULTURE OFFERS NOT ONLY A SOLUTION TO THE CHALLENGES FACING MODERN AGRICULTURE BUT ALSO A GLIMPSE INTO A MORE HARMONIOUS RELATIONSHIP BETWEEN HUMANITY AND THE NATURAL WORLD.

AS WE DIVE DEEPER INTO THE INTRICACIES OF ELECTROCULTURE, MAY YOU BE INSPIRED TO EMBRACE THESE INNOVATIVE TECHNOLOGIES AND JOIN THE MOVEMENT TOWARDS A GREENER, MORE SUSTAINABLE FUTURE.

JOIN US ON THIS JOURNEY OF DISCOVERY, INNOVATION, AND TRANSFORMATION. TOGETHER, WE WILL EXPLORE THE BOUNDLESS POTENTIAL OF ADVANCED ELECTROCULTURE AND

DAVID HOLMAN

PAVE THE WAY FOR A BRIGHTER, MORE SUSTAINABLE TOMORROW.

WELCOME TO THE WORLD OF ADVANCED ELECTROCULTURE.

CHAPTER 1

UNDERSTANDING ELECTROCULTURE BASICS

IN THE WORLD OF AGRICULTURE, WHERE TRADITIONAL PRACTICES HAVE BEEN THE NORM FOR GENERATIONS, ELECTROCULTURE EMERGES AS A TRANSFORMATIVE FORCE. IT PROMISES TO CHALLENGE ESTABLISHED CONVENTIONS, OFFERING NEW POSSIBILITIES FOR ENHANCING CROP YIELDS, IMPROVING SOIL HEALTH, AND REDUCING THE ENVIRONMENTAL IMPACT OF FARMING. IN THIS CHAPTER, WE EMBARK ON A JOURNEY THROUGH THE ANNALS OF ELECTROCULTURE, EXPLORING ITS HISTORICAL CONTEXT, THE SERENDIPITOUS DISCOVERIES THAT LED TO ITS DEVELOPMENT,

DAVID HOLMAN

AND THE FASCINATING CONNECTION BETWEEN ELECTRICITY AND PLANT GROWTH.

THE HISTORICAL CONTEXT

TO FULLY GRASP THE SIGNIFICANCE OF ELECTROCULTURE, WE MUST FIRST STEP BACK IN TIME AND DELVE INTO ITS HISTORICAL ROOTS. THE STORY OF ELECTROCULTURE IS ONE THAT INTERTWINES THE CURIOSITY OF EARLY EXPERIMENTERS, THE MYSTERIES OF PLANT PHYSIOLOGY, AND THE RELENTLESS PURSUIT OF SUSTAINABLE AGRICULTURAL PRACTICES.

ANCIENT BEGINNINGS

THE ORIGINS OF ELECTROCULTURE CAN BE TRACED BACK TO ANCIENT CIVILIZATIONS, WHERE EARLY FARMERS

OBSERVED PECULIAR PHENOMENA RELATED TO PLANT GROWTH. IN REGIONS SUCH AS MESOPOTAMIA AND THE NILE VALLEY, WHERE THE FOUNDATIONS OF AGRICULTURE WERE LAID, ANECDOTAL EVIDENCE SUGGESTS THAT CERTAIN PLANTS SEEMED TO FLOURISH NEAR NATURAL SOURCES OF ELECTRICITY, SUCH AS LIGHTNING STRIKES.

ONE OF THE EARLIEST RECORDED ACCOUNTS OF ELECTROCULTURE DATES BACK TO THE ANCIENT GREEKS AND THEIR USE OF A MINERAL CALLED "TOURMALINE." KNOWN FOR ITS PYROELECTRIC PROPERTIES, TOURMALINE COULD GENERATE SMALL ELECTRICAL CHARGES WHEN HEATED OR RUBBED. IT IS SAID THAT THE GREEKS USED POWDERED TOURMALINE AS A FERTILIZER, BELIEVING THAT THE ELECTRICAL ENERGY IT RELEASED HAD A

POSITIVE EFFECT ON PLANT GROWTH.

WHILE THESE EARLY PRACTICES LACKED SCIENTIFIC VALIDATION, THEY PLANTED THE SEEDS OF CURIOSITY THAT WOULD LATER FLOURISH INTO THE FIELD OF ELECTROCULTURE.

THE AGE OF ENLIGHTENMENT

THE 18TH CENTURY MARKED A PERIOD OF SIGNIFICANT SCIENTIFIC EXPLORATION AND EXPERIMENTATION. DURING THE AGE OF ENLIGHTENMENT, SCHOLARS AND NATURALISTS BEGAN TO DISSECT THE MYSTERIES OF ELECTRICITY AND ITS POTENTIAL APPLICATIONS IN VARIOUS FIELDS, INCLUDING AGRICULTURE.

ONE OF THE KEY FIGURES IN THIS ERA WAS LUIGI GALVANI, AN ITALIAN PHYSICIAN AND PHYSICIST. IN 1780, GALVANI

CONDUCTED EXPERIMENTS INVOLVING THE LEGS OF

DISSECTED FROGS AND DISCOVERED THAT THEY TWITCHED WHEN EXPOSED TO ELECTRICAL SPARKS. THIS OBSERVATION LED TO THE CONCEPT OF "ANIMAL

DAVID HOLMAN

ELECTRICITY" AND IGNITED A SCIENTIFIC FASCINATION WITH THE RELATIONSHIP BETWEEN ELECTRICITY AND LIVING ORGANISMS.

GALVANI'S WORK LAID THE FOUNDATION FOR FURTHER RESEARCH INTO THE EFFECTS OF ELECTRICITY ON BIOLOGICAL SYSTEMS, INCLUDING PLANTS. HIS NEPHEW, GIOVANNI ALDINI, EXTENDED THESE INVESTIGATIONS BY APPLYING ELECTRICAL CURRENTS TO PLANT TISSUES, PROMPTING THEM TO EXHIBIT VARIOUS RESPONSES. ALTHOUGH THEIR EXPERIMENTS WERE RUDIMENTARY BY MODERN STANDARDS, THEY MARKED A SIGNIFICANT STEP TOWARD UNDERSTANDING THE POTENTIAL INFLUENCE OF ELECTRICITY ON PLANT LIFE.

THE EMERGENCE OF ELECTROCULTURE

THE TERM "ELECTROCULTURE" ITSELF CAN BE ATTRIBUTED TO THE FRENCH PHYSIOLOGIST AND BIOLOGIST RENÉ ANTOINE FERCHAULT DE RÉAUMUR. IN THE 18TH CENTURY, RÉAUMUR CONDUCTED EXPERIMENTS INVOLVING THE ELECTRIFICATION OF SOIL, SEEDS, AND PLANTS. HE OBSERVED THAT ELECTRIFIED SOIL APPEARED TO PROMOTE HEALTHIER PLANT GROWTH AND GREATER CROP YIELDS.

RÉAUMUR'S PIONEERING WORK LAID THE CONCEPTUAL GROUNDWORK FOR ELECTROCULTURE, ALTHOUGH IT REMAINED A RELATIVELY OBSCURE FIELD AT THE TIME. HIS FINDINGS WERE MET WITH SKEPTICISM, AS THEY CHALLENGED THE PREVAILING AGRICULTURAL PRACTICES OF THE DAY.

DAVID HOLMAN

THE ADVENT OF MODERN ELECTROCULTURE

THE 19TH AND 20TH CENTURIES WITNESSED A RESURGENCE OF INTEREST IN ELECTROCULTURE, DRIVEN BY ADVANCES IN SCIENCE AND TECHNOLOGY. AS ELECTRICITY BECAME MORE WIDELY UNDERSTOOD AND HARNESSED, RESEARCHERS AND AGRICULTURALISTS BEGAN TO REVISIT THE POTENTIAL APPLICATIONS OF ELECTRICITY IN FARMING.

ONE NOTABLE FIGURE IN THIS RESURGENCE WAS GEORGES LAKHOVSKY, A RUSSIAN-BORN ENGINEER AND INVENTOR. LAKHOVSKY DEVELOPED THE MULTIPLE WAVE OSCILLATOR, A DEVICE THAT GENERATED A RANGE OF ELECTROMAGNETIC FREQUENCIES. HE BELIEVED THAT THESE FREQUENCIES COULD STIMULATE THE GROWTH AND

HEALTH OF PLANTS. LAKHOVSKY'S WORK GAINED ATTENTION, AND HE CONDUCTED EXPERIMENTS IN VARIOUS COUNTRIES TO DEMONSTRATE THE POTENTIAL OF ELECTROMAGNETIC FIELDS IN AGRICULTURE.

DESPITE THE INTRIGUING RESULTS OBTAINED BY EARLY ELECTROCULTURE PIONEERS LIKE LAKHOVSKY, THE SCIENTIFIC COMMUNITY REMAINED DIVIDED. MANY DISMISSED ELECTROCULTURE AS PSEUDOSCIENCE, WHILE OTHERS CONTINUED TO EXPLORE ITS POSSIBILITIES.

THE CONTEMPORARY ELECTROCULTURE MOVEMENT

IN RECENT DECADES, ELECTROCULTURE HAS EXPERIENCED A REVIVAL, DRIVEN BY ADVANCES IN SCIENTIFIC UNDERSTANDING AND

TECHNOLOGY. RESEARCHERS AROUND THE WORLD HAVE REVISITED THE CONCEPT, APPLYING MODERN SCIENTIFIC METHODOLOGIES TO UNRAVEL THE MYSTERIES OF ELECTRICITY'S INFLUENCE ON PLANTS.

ONE CONTEMPORARY PROPONENT OF ELECTROCULTURE IS DR. T. GALEN HIERONYMUS, AN AMERICAN SCIENTIST WHO HAS DEVOTED HIS CAREER TO THE STUDY OF SUBTLE ENERGIES AND THEIR EFFECTS ON AGRICULTURE. HIS RESEARCH HAS FOCUSED ON THE USE OF RADIONICS AND OTHER ENERGY-BASED TECHNOLOGIES TO ENHANCE CROP GROWTH AND SOIL HEALTH. DR. HIERONYMUS' WORK UNDERSCORES THE ENDURING INTEREST IN ELECTROCULTURE AS A MEANS TO IMPROVE AGRICULTURAL PRACTICES.

DAVID HOLMAN

ELECTROCULTURE IN PRACTICE TODAY

TODAY, ELECTROCULTURE STANDS AT THE INTERSECTION OF TRADITION AND INNOVATION. IT DRAWS FROM CENTURIES OF ANECDOTAL EVIDENCE, EARLY EXPERIMENTATION, AND CONTEMPORARY SCIENTIFIC INQUIRY TO OFFER A NEW PERSPECTIVE ON AGRICULTURE.

WHILE ELECTROCULTURE REMAINS A FIELD OF ONGOING RESEARCH AND EXPLORATION, SOME FARMERS AND AGRICULTURALISTS HAVE BEGUN TO ADOPT ITS PRINCIPLES. THEY EXPERIMENT WITH VARIOUS FORMS OF ELECTRICAL STIMULATION, ELECTROMAGNETIC FIELDS, AND OTHER TECHNOLOGIES TO ENHANCE PLANT GROWTH, INCREASE CROP YIELDS, AND REDUCE THE NEED FOR CHEMICAL INPUTS.

DAVID HOLMAN

THE RESURGENCE OF INTEREST IN ELECTROCULTURE IS DRIVEN BY A

GROWING AWARENESS OF THE ENVIRONMENTAL CHALLENGES FACING MODERN AGRICULTURE. SOIL DEGRADATION, WATER POLLUTION, AND THE DEPLETION OF NATURAL RESOURCES HAVE

DAVID HOLMAN

LED TO A SEARCH FOR MORE SUSTAINABLE FARMING PRACTICES. ELECTROCULTURE, WITH ITS POTENTIAL TO IMPROVE SOIL HEALTH AND BOOST CROP RESILIENCE, HAS EMERGED AS A PROMISING AVENUE FOR SUSTAINABLE AGRICULTURE.

THE ELECTRIC CONNECTION TO PLANT GROWTH

AT THE HEART OF ELECTROCULTURE LIES THE INTRIGUING RELATIONSHIP BETWEEN ELECTRICITY AND PLANT GROWTH. TO UNDERSTAND THIS CONNECTION, WE MUST EXPLORE THE WAYS IN WHICH ELECTRICAL CURRENTS AND FIELDS INTERACT WITH PLANT PHYSIOLOGY.

ELECTROTROPISM AND ELECTROKINESIS

PLANTS, LIKE ALL LIVING ORGANISMS, ARE COMPOSED OF

CELLS THAT CARRY ELECTRICAL CHARGES. THESE CHARGES PLAY A FUNDAMENTAL ROLE IN VARIOUS PHYSIOLOGICAL PROCESSES. TWO KEY PHENOMENA THAT ILLUSTRATE THE PLANT-ELECTRICITY CONNECTION ARE ELECTROTROPISM AND ELECTROKINESIS.

ELECTROTROPISM REFERS TO THE DIRECTIONAL GROWTH OF PLANTS IN RESPONSE TO ELECTRIC FIELDS. JUST AS PLANTS EXHIBIT PHOTOTROPISM BY GROWING TOWARD LIGHT SOURCES, THEY CAN ALSO RESPOND TO ELECTRICAL GRADIENTS IN THEIR ENVIRONMENT. WHEN EXPOSED TO AN ELECTRIC FIELD, THE TIPS OF PLANT ROOTS TEND TO GROW TOWARD THE POSITIVE ELECTRODE, WHILE THE SHOOTS GROW TOWARD THE NEGATIVE ELECTRODE. THIS PHENOMENON HAS BEEN OBSERVED IN VARIOUS

PLANT SPECIES AND SUGGESTS THAT PLANTS CAN SENSE AND RESPOND TO ELECTRICAL CUES IN THEIR SURROUNDINGS.

ELECTROKINESIS, ON THE OTHER HAND, INVOLVES CHANGES IN PLANT CELL MOVEMENT IN RESPONSE TO ELECTRICAL STIMULATION. IT HAS BEEN OBSERVED THAT ELECTRICAL CURRENTS CAN INFLUENCE THE OPENING AND CLOSING OF STOMATA (SMALL OPENINGS ON PLANT SURFACES), AS WELL AS THE MOVEMENT OF CELL ORGANELLES. THESE SUBTLE RESPONSES TO ELECTRICAL STIMULI HINT AT THE INTRICATE WAYS IN WHICH PLANTS CAN PERCEIVE AND INTERACT WITH ELECTRICAL FORCES.

ELECTROCHEMICAL SIGNALING

BENEATH THE SURFACE OF PLANT TISSUES, A COMPLEX NETWORK OF ELECTROCHEMICAL SIGNALING PATHWAYS OPERATES. PLANTS

USE ELECTRICAL SIGNALS TO COORDINATE GROWTH, RESPOND TO ENVIRONMENTAL CUES, AND DEFEND AGAINST STRESSORS SUCH AS HERBIVORES AND PATHOGENS.

DAVID HOLMAN

ONE REMARKABLE EXAMPLE OF ELECTROCHEMICAL SIGNALING IS THE PROPAGATION OF ACTION POTENTIALS IN PLANTS. SIMILAR TO THE WAY NERVE CELLS TRANSMIT SIGNALS IN ANIMALS, PLANTS CAN GENERATE AND TRANSMIT ELECTRICAL IMPULSES ALONG THEIR CELLS. THESE IMPULSES CAN TRAVEL RAPIDLY, ENABLING PLANTS TO RESPOND TO EXTERNAL STIMULI WITH IMPRESSIVE SPEED.

FOR INSTANCE, WHEN A LEAF IS WOUNDED, A RAPID ELECTRICAL SIGNAL CAN BE TRANSMITTED FROM THE INJURED SITE TO OTHER PARTS OF THE PLANT, TRIGGERING A SERIES OF DEFENSE RESPONSES. THESE RESPONSES MAY INCLUDE THE PRODUCTION OF DEFENSIVE CHEMICALS, THE CLOSURE OF STOMATA TO REDUCE WATER LOSS, AND THE REINFORCEMENT

DAVID HOLMAN

OF CELL WALLS TO RESIST
HERBIVORE DAMAGE.

NUTRIENT UPTAKE AND ELECTRICAL CONDUCTIVITY

ELECTRICAL CONDUCTIVITY, A PROPERTY RELATED TO THE ABILITY OF MATERIALS TO CONDUCT ELECTRICAL CURRENTS, ALSO PLAYS A ROLE IN PLANT HEALTH. SOIL ELECTRICAL CONDUCTIVITY INFLUENCES THE AVAILABILITY OF NUTRIENTS TO PLANTS. SOILS WITH HIGHER ELECTRICAL CONDUCTIVITY TEND TO HAVE BETTER NUTRIENT AVAILABILITY, AS THE ELECTRICAL CURRENTS FACILITATE THE MOVEMENT OF IONS (CHARGED PARTICLES) IN THE SOIL SOLUTION. THIS IS PARTICULARLY IMPORTANT FOR ESSENTIAL NUTRIENTS SUCH AS POTASSIUM, CALCIUM, AND MAGNESIUM, WHICH ARE TAKEN UP BY PLANT ROOTS AS IONS.

DAVID HOLMAN

IN ELECTROCULTURE, THE APPLICATION OF ELECTRICAL CURRENTS TO SOIL CAN INFLUENCE ITS ELECTRICAL CONDUCTIVITY AND, IN TURN, NUTRIENT AVAILABILITY. BY STRATEGICALLY MANAGING SOIL CONDUCTIVITY, FARMERS CAN OPTIMIZE NUTRIENT UPTAKE BY CROPS, POTENTIALLY LEADING TO IMPROVED GROWTH AND YIELDS.

ELECTROCULTURE AND PHOTOSYNTHESIS

PHOTOSYNTHESIS, THE PROCESS BY WHICH PLANTS CONVERT LIGHT ENERGY INTO CHEMICAL ENERGY, IS CENTRAL TO THEIR GROWTH AND SURVIVAL. IT TURNS OUT THAT ELECTRICAL FACTORS ALSO INFLUENCE PHOTOSYNTHESIS. IN PARTICULAR, THE ELECTRICAL POTENTIAL (OR VOLTAGE) ACROSS THE THYLAKOID MEMBRANE WITHIN CHLOROPLASTS, THE

CELLULAR STRUCTURES RESPONSIBLE FOR PHOTOSYNTHESIS, PLAYS A CRUCIAL ROLE.

THE THYLAKOID MEMBRANE'S ELECTRICAL POTENTIAL IS INVOLVED IN THE GENERATION OF ADENOSINE TRIPHOSPHATE (ATP), A MOLECULE THAT STORES AND TRANSFERS ENERGY WITHIN CELLS. ATP IS A KEY PLAYER IN THE PHOTOSYNTHETIC PROCESS,

DAVID HOLMAN

POWERING THE CONVERSION OF
CARBON DIOXIDE AND WATER
INTO GLUCOSE AND OXYGEN.

ELECTROCULTURE RESEARCHERS
ARE EXPLORING WAYS TO
OPTIMIZE THE ELECTRICAL
CONDITIONS WITHIN PLANT CELLS
TO ENHANCE PHOTOSYNTHESIS.
BY DOING SO, THEY AIM TO
IMPROVE CROP PRODUCTIVITY
AND MAKE AGRICULTURE MORE
EFFICIENT.

THE JOURNEY INTO THE WORLD
OF ELECTROCULTURE HAS BEGUN
WITH A LOOK AT ITS HISTORICAL
ROOTS AND THE CAPTIVATING
CONNECTION BETWEEN
ELECTRICITY AND PLANT
GROWTH. FROM ANCIENT
CIVILIZATIONS TO
CONTEMPORARY SCIENTISTS, THE
FASCINATION WITH
ELECTROCULTURE HAS
PERSISTED, DRIVEN BY THE HOPE
OF TRANSFORMING AGRICULTURE

DAVID HOLMAN

INTO A MORE SUSTAINABLE AND
PRODUCTIVE ENDEAVOR.

AS WE VENTURE FURTHER INTO
THE REALM OF ELECTROCULTURE,
WE WILL EXPLORE THE PRACTICAL
APPLICATIONS, INNOVATIVE
TECHNOLOGIES, AND REAL-WORLD
SUCCESS STORIES THAT DEFINE
THIS FIELD TODAY. THE CHAPTERS
THAT FOLLOW WILL UNRAVEL THE
MYSTERIES OF HARNESSING
ELECTRICITY TO BOOST CROP
YIELDS, IMPROVE SOIL HEALTH,
AND REDUCE THE
ENVIRONMENTAL FOOTPRINT OF
FARMING.

DAVID HOLMAN

CHAPTER 2

EARTH BATTERIES IN AGRICULTURE

IN THE WORLD OF AGRICULTURE, INNOVATION OFTEN EMERGES FROM UNEXPECTED SOURCES. IN

DAVID HOLMAN

THE CASE OF ELECTROCULTURE, ONE SUCH UNEXPECTED SOURCE OF SUSTAINABLE ENERGY IS EARTH BATTERIES. THESE UNASSUMING YET POWERFUL DEVICES HARNESS THE EARTH'S NATURAL ELECTRICAL POTENTIAL TO PROVIDE A RENEWABLE SOURCE OF ENERGY FOR AGRICULTURAL APPLICATIONS. IN THIS CHAPTER, WE WILL EXPLORE THE PRINCIPLES BEHIND EARTH BATTERIES, THEIR HISTORICAL CONTEXT, AND THEIR REMARKABLE CAPACITY TO TRANSFORM FARMING PRACTICES.

EARTH BATTERIES DEFINED

BEFORE WE DELVE INTO THE PRACTICAL APPLICATIONS OF EARTH BATTERIES IN AGRICULTURE, IT IS ESSENTIAL TO UNDERSTAND THE BASIC PRINCIPLES THAT UNDERLIE THESE REMARKABLE DEVICES.

THE PRINCIPLE OF GALVANIC CELLS

AT THE HEART OF AN EARTH BATTERY LIES THE PRINCIPLE OF GALVANIC CELLS, A CONCEPT THAT DATES BACK TO THE 18TH CENTURY. GALVANIC CELLS, ALSO KNOWN AS VOLTAIC CELLS OR ELECTROCHEMICAL CELLS, ARE DEVICES THAT CONVERT CHEMICAL ENERGY INTO ELECTRICAL ENERGY THROUGH CHEMICAL REACTIONS.

IN A TYPICAL GALVANIC CELL, TWO DIFFERENT METALS OR MATERIALS ARE IMMERSED IN AN ELECTROLYTE SOLUTION, WHICH ALLOWS IONS TO MOVE BETWEEN THEM. THIS MOVEMENT OF IONS CREATES A FLOW OF ELECTRONS, RESULTING IN AN ELECTRIC CURRENT. THIS ELECTRIC CURRENT CAN BE HARNESSED TO POWER VARIOUS ELECTRICAL DEVICES.

DAVID HOLMAN

EARTH BATTERIES ARE A SPECIFIC TYPE OF GALVANIC CELL THAT UTILIZES THE EARTH ITSELF AS PART OF THE CELL'S COMPONENTS. THEY TYPICALLY CONSIST OF THREE KEY ELEMENTS:

1. **ANODE:** THIS IS THE ELECTRODE THAT UNDERGOES OXIDATION, RELEASING ELECTRONS INTO THE CIRCUIT. IN AN EARTH BATTERY, THE ANODE IS USUALLY A METAL ROD OR ELECTRODE BURIED IN THE GROUND.

2. **CATHODE:** THIS IS THE ELECTRODE WHERE REDUCTION TAKES PLACE, CONSUMING ELECTRONS FROM THE CIRCUIT. THE CATHODE CAN BE ANOTHER METAL ELECTRODE OR A MATERIAL THAT REACTS WITH THE ANODE'S ELECTRONS.

3. **ELECTROLYTE:** THE ELECTROLYTE IS THE MEDIUM THROUGH WHICH IONS CAN MOVE BETWEEN THE ANODE AND CATHODE. IN EARTH BATTERIES, THE EARTH ITSELF SERVES AS THE ELECTROLYTE, ALLOWING IONS TO FLOW BETWEEN THE BURIED ANODE AND CATHODE.

THE CHEMICAL REACTIONS THAT OCCUR AT THE ANODE AND CATHODE GENERATE AN ELECTRIC POTENTIAL DIFFERENCE, OR VOLTAGE, BETWEEN THE TWO ELECTRODES. THIS VOLTAGE CAN BE TAPPED INTO TO POWER ELECTRICAL DEVICES OR SYSTEMS.

HARNESSING EARTH'S POTENTIAL

EARTH BATTERIES DERIVE THEIR NAME FROM THE FACT THAT THEY UTILIZE THE EARTH'S NATURAL

ELECTRICAL POTENTIAL TO CREATE A VOLTAGE DIFFERENCE. THE EARTH'S SURFACE IS NOT ELECTRICALLY NEUTRAL; RATHER, IT CARRIES A SLIGHT NEGATIVE CHARGE DUE TO THE PRESENCE OF EXCESS ELECTRONS. THIS NEGATIVE CHARGE CAN SERVE AS THE CATHODE OF AN EARTH BATTERY.

WHEN A METAL ROD OR ELECTRODE IS INSERTED INTO THE GROUND, IT BECOMES THE ANODE OF THE EARTH BATTERY. THE CHEMICAL REACTIONS THAT OCCUR AT THE ANODE AND CATHODE RESULT IN THE FLOW OF ELECTRONS THROUGH AN EXTERNAL CIRCUIT, CREATING A USABLE ELECTRIC CURRENT.

EARTH BATTERIES ARE OFTEN DESCRIBED AS "SELF-RENEWING" BECAUSE THE EARTH'S ELECTRICAL POTENTIAL IS CONTINUOUSLY REPLENISHED BY

DAVID HOLMAN

NATURAL PROCESSES, SUCH AS LIGHTNING STRIKES AND THE EARTH'S INTERACTION WITH THE IONOSPHERE. THIS MEANS THAT EARTH BATTERIES CAN PROVIDE A CONSISTENT SOURCE OF RENEWABLE ENERGY AS LONG AS THE CHEMICAL REACTIONS AT THE ELECTRODES CONTINUE.

HISTORICAL PERSPECTIVES

THE USE OF EARTH BATTERIES IN AGRICULTURE MAY SEEM LIKE A MODERN INNOVATION, BUT ITS ROOTS EXTEND DEEP INTO HISTORY. TO APPRECIATE THE SIGNIFICANCE OF EARTH BATTERIES IN CONTEMPORARY FARMING, IT IS ESSENTIAL TO EXPLORE THEIR HISTORICAL CONTEXT.

EARLY EXPERIMENTS

THE IDEA OF UTILIZING THE EARTH'S ELECTRICAL POTENTIAL FOR PRACTICAL PURPOSES CAN BE

TRACED BACK TO THE LATE 18TH CENTURY. SCIENTISTS AND INVENTORS OF THAT ERA WERE CAPTIVATED BY THE POSSIBILITIES OF ELECTRICITY, AND THEY CONDUCTED A RANGE OF EXPERIMENTS TO HARNESS ITS POWER.

ONE OF THE EARLY EXPERIMENTS THAT LAID THE FOUNDATION FOR EARTH BATTERIES WAS CONDUCTED BY LUIGI GALVANI, THE SAME ITALIAN PHYSICIAN AND PHYSICIST KNOWN FOR HIS WORK ON ANIMAL ELECTRICITY. IN THE LATE 1780S, GALVANI'S NEPHEW, GIOVANNI ALDINI, EXPERIMENTED WITH THE ELECTRICAL PROPERTIES OF METALS AND THEIR INTERACTION WITH BIOLOGICAL TISSUES. HIS WORK, WHILE FOCUSED ON ANIMAL PHYSIOLOGY, CONTRIBUTED TO THE UNDERSTANDING OF THE ELECTRICAL BEHAVIOR OF METALS

DAVID HOLMAN

IN DIFFERENT ENVIRONMENTS, INCLUDING THE EARTH.

EARTH BATTERIES IN THE 19TH CENTURY

THE 19TH CENTURY WITNESSED A SURGE IN EXPERIMENTATION WITH GALVANIC CELLS AND BATTERIES. RESEARCHERS AND INVENTORS EXPLORED VARIOUS CONFIGURATIONS, MATERIALS, AND APPLICATIONS. EARTH BATTERIES, WHICH UTILIZED READILY AVAILABLE MATERIALS AND HARNESSED THE EARTH'S ELECTRICAL POTENTIAL, GAINED ATTENTION DURING THIS PERIOD.

IN 1836, CHARLES GRAFTON PAGE, AN AMERICAN SCIENTIST AND INVENTOR, PATENTED AN "IMPROVED METHOD OF DISTRIBUTING ELECTRICITY." HIS INVENTION INVOLVED USING EARTH BATTERIES TO POWER ELECTRICAL DEVICES AT A

DISTANCE. PAGE'S WORK
FORESHADOWED THE CONCEPT OF
TRANSMITTING ELECTRICAL
ENERGY OVER LONG DISTANCES, A
NOTION THAT WOULD LATER
BECOME A REALITY WITH THE
ADVENT OF ALTERNATING
CURRENT (AC) POWER
TRANSMISSION.

DURING THE SAME ERA, EARTH
BATTERIES FOUND APPLICATIONS
IN TELEGRAPHY. THE TELEGRAPH,
A REVOLUTIONARY MEANS OF
LONG-DISTANCE COMMUNICATION,
RELIED ON BATTERIES TO
TRANSMIT ELECTRICAL SIGNALS
OVER TELEGRAPH WIRES. EARTH
BATTERIES OFFERED A RELIABLE
AND ECONOMICAL SOURCE OF
POWER FOR TELEGRAPH
STATIONS, PARTICULARLY IN
REMOTE AREAS.

AGRICULTURE AND EARTH BATTERIES

WHILE THE HISTORICAL RECORD PROVIDES EVIDENCE OF EARTH BATTERIES BEING USED FOR TELEGRAPHY AND SCIENTIFIC EXPERIMENTATION, THEIR APPLICATION IN AGRICULTURE REMAINED RELATIVELY UNEXPLORED UNTIL RECENT DECADES. EARLY AGRICULTURAL PRACTICES WERE DEEPLY ROOTED IN TRADITIONAL METHODS, AND THE ADOPTION OF NEW TECHNOLOGIES WAS GRADUAL.

IN THE 20TH CENTURY, AGRICULTURE UNDERWENT SIGNIFICANT TRANSFORMATIONS WITH THE INTRODUCTION OF MECHANIZATION, SYNTHETIC FERTILIZERS, AND PESTICIDES. THESE ADVANCES LED TO INCREASED CROP YIELDS BUT ALSO RAISED CONCERNS ABOUT SOIL DEGRADATION AND ENVIRONMENTAL IMPACT. THE QUEST FOR SUSTAINABLE

DAVID HOLMAN

FARMING PRACTICES REIGNITED INTEREST IN UNCONVENTIONAL SOURCES OF ENERGY, SUCH AS EARTH BATTERIES.

HARNESSING EARTH BATTERIES FOR FARMING

THE APPLICATION OF EARTH BATTERIES IN AGRICULTURE REPRESENTS A MODERN-DAY ADAPTATION OF HISTORICAL PRINCIPLES. FARMERS AND RESEARCHERS HAVE RECOGNIZED THE POTENTIAL OF EARTH BATTERIES TO PROVIDE SUSTAINABLE AND RENEWABLE ENERGY FOR VARIOUS FARMING OPERATIONS. IN THIS SECTION, WE WILL EXPLORE HOW EARTH BATTERIES ARE HARNESSED FOR PRACTICAL USE IN AGRICULTURE.

SOIL ELECTROCHEMISTRY

TO UNDERSTAND HOW EARTH BATTERIES CAN BE USED IN AGRICULTURE, IT IS ESSENTIAL TO

GRASP THE PRINCIPLES OF SOIL ELECTROCHEMISTRY. SOIL, LIKE ANY NATURAL MATERIAL, HAS ELECTRICAL PROPERTIES THAT INFLUENCE ITS BEHAVIOR. THESE PROPERTIES INCLUDE ELECTRICAL CONDUCTIVITY, ION EXCHANGE CAPACITY, AND REDOX POTENTIAL.

1. **ELECTRICAL CONDUCTIVITY:** SOIL ELECTRICAL CONDUCTIVITY IS A MEASURE OF ITS ABILITY TO CONDUCT ELECTRICAL CURRENTS. IT DEPENDS ON FACTORS SUCH AS SOIL MOISTURE, ION CONCENTRATION, AND THE TYPES OF IONS PRESENT. SOILS WITH HIGHER ELECTRICAL CONDUCTIVITY ALLOW ELECTRICAL CURRENTS TO FLOW MORE EASILY.

2. **ION EXCHANGE CAPACITY (CEC):** SOIL HAS THE CAPACITY

TO EXCHANGE IONS WITH THE SURROUNDING ENVIRONMENT. THIS PROPERTY, KNOWN AS CATION EXCHANGE CAPACITY (CEC), INFLUENCES THE AVAILABILITY OF NUTRIENTS TO PLANTS. SOILS WITH HIGHER CEC CAN RETAIN AND RELEASE MORE IONS, AFFECTING NUTRIENT UPTAKE BY CROPS.

3.   **REDOX POTENTIAL:** SOIL REDOX POTENTIAL IS A MEASURE OF ITS OXIDATIVE OR REDUCING PROPERTIES. IT INDICATES THE PRESENCE OF OXYGEN (OXIDIZING CONDITIONS) OR THE ABSENCE OF OXYGEN (REDUCING CONDITIONS) IN THE SOIL. REDOX POTENTIAL INFLUENCES NUTRIENT AVAILABILITY, MICROBIAL ACTIVITY, AND SOIL HEALTH.

EARTH BATTERIES LEVERAGE THESE SOIL PROPERTIES TO

CREATE A VOLTAGE DIFFERENCE BETWEEN THE BURIED ANODE AND THE EARTH ITSELF. THIS VOLTAGE DIFFERENCE CAN BE HARNESSED TO POWER VARIOUS ELECTRICAL DEVICES OR SYSTEMS.

PRACTICAL APPLICATIONS IN AGRICULTURE

EARTH BATTERIES FIND PRACTICAL APPLICATIONS IN AGRICULTURE ACROSS A RANGE OF SCENARIOS. THEIR VERSATILITY AND RENEWABLE NATURE MAKE THEM VALUABLE ASSETS FOR FARMERS SEEKING SUSTAINABLE SOLUTIONS. LET'S EXPLORE SOME OF THE KEY APPLICATIONS:

1. **REMOTE MONITORING AND DATA COLLECTION:** IN MODERN AGRICULTURE, REMOTE SENSORS AND DATA COLLECTION SYSTEMS ARE ESSENTIAL FOR MONITORING

SOIL CONDITIONS, WEATHER, AND CROP HEALTH. EARTH BATTERIES PROVIDE A RELIABLE SOURCE OF POWER FOR THESE SENSORS, ELIMINATING THE NEED FOR FREQUENT BATTERY REPLACEMENT OR RELIANCE ON GRID ELECTRICITY IN REMOTE LOCATIONS.

2. **ELECTRIC FENCING:** ELECTRIC FENCING IS A COMMON METHOD USED BY FARMERS TO PROTECT CROPS FROM WILDLIFE AND DETER LIVESTOCK FROM STRAYING. EARTH BATTERIES CAN POWER ELECTRIC FENCING SYSTEMS, PROVIDING A COST-EFFECTIVE AND ENVIRONMENTALLY FRIENDLY SOLUTION.

3. **IRRIGATION SYSTEMS:** SUSTAINABLE IRRIGATION PRACTICES, SUCH AS DRIP IRRIGATION, BENEFIT FROM

EARTH BATTERIES TO POWER
CONTROL SYSTEMS AND
SENSORS. THIS ENSURES
PRECISE WATER MANAGEMENT,
REDUCING WATER WASTAGE
AND OPTIMIZING CROP
HYDRATION.

4. **SOIL HEALTH
MONITORING:** SOIL HEALTH IS
A CRITICAL FACTOR IN
AGRICULTURE, AND
CONTINUOUS MONITORING IS
ESSENTIAL. EARTH BATTERIES
CAN POWER SOIL SENSORS
THAT MEASURE KEY
PARAMETERS LIKE MOISTURE
CONTENT, PH, AND NUTRIENT
LEVELS, ALLOWING FARMERS
TO MAKE DATA-DRIVEN
DECISIONS FOR SOIL
IMPROVEMENT.

5. **GREENHOUSES AND
CONTROLLED
ENVIRONMENT
AGRICULTURE:** EARTH

BATTERIES CAN CONTRIBUTE TO THE ENERGY NEEDS OF GREENHOUSE OPERATIONS, WHERE MAINTAINING OPTIMAL CONDITIONS FOR PLANT GROWTH IS CRUCIAL. THEY CAN POWER VENTILATION SYSTEMS, CLIMATE CONTROL, AND LIGHTING.

6.   **ELECTROCULTURE SYSTEMS:** EARTH BATTERIES CAN PLAY A PIVOTAL ROLE IN ELECTROCULTURE SYSTEMS, WHERE THEY PROVIDE THE NECESSARY ELECTRICAL ENERGY TO STIMULATE PLANT GROWTH AND NUTRIENT UPTAKE. THE INTEGRATION OF EARTH BATTERIES WITH ELECTROCULTURE HOLDS SIGNIFICANT POTENTIAL FOR SUSTAINABLE CROP PRODUCTION.

IMPLEMENTING EARTH BATTERIES

THE IMPLEMENTATION OF EARTH BATTERIES IN AGRICULTURE INVOLVES CAREFUL PLANNING AND DESIGN. SEVERAL FACTORS MUST BE CONSIDERED TO MAXIMIZE THE EFFICIENCY AND EFFECTIVENESS OF THESE SYSTEMS:

1. **SITE SELECTION:** THE LOCATION OF THE EARTH BATTERY IS CRITICAL. IT SHOULD BE CHOSEN BASED ON FACTORS SUCH AS SOIL CONDUCTIVITY, MOISTURE LEVELS, AND ACCESSIBILITY. CONDUCTING A SOIL ANALYSIS CAN HELP IDENTIFY SUITABLE LOCATIONS.

2. **ELECTRODE MATERIALS:** THE CHOICE OF MATERIALS FOR THE ANODE AND CATHODE ELECTRODES AFFECTS THE PERFORMANCE OF THE EARTH BATTERY. COMMON ELECTRODE MATERIALS INCLUDE COPPER,

IRON, AND ZINC. THE COMPATIBILITY OF ELECTRODE MATERIALS WITH THE SOIL AND THEIR LONGEVITY SHOULD BE CONSIDERED.

3. **ELECTRODE CONFIGURATION:** THE ARRANGEMENT OF ELECTRODES, SPACING, AND DEPTH OF BURIAL ARE ESSENTIAL DESIGN CONSIDERATIONS. THESE FACTORS INFLUENCE THE ELECTRICAL POTENTIAL GENERATED BY THE EARTH BATTERY.

4. **LOAD REQUIREMENTS:** UNDERSTANDING THE ENERGY REQUIREMENTS OF THE INTENDED APPLICATIONS IS CRUCIAL. PROPER SIZING OF THE EARTH BATTERY SYSTEM ENSURES THAT IT CAN MEET THE DEMANDS OF THE ELECTRICAL LOAD.

5. MONITORING AND MAINTENANCE: REGULAR

MONITORING OF THE EARTH BATTERY SYSTEM IS NECESSARY TO ENSURE ITS CONTINUED OPERATION. MAINTENANCE MAY INVOLVE CLEANING ELECTRODES, CHECKING ELECTRICAL CONNECTIONS, AND MONITORING SOIL CONDITIONS.

DAVID HOLMAN

6. **SAFETY CONSIDERATIONS:** EARTH BATTERY SYSTEMS SHOULD BE DESIGNED AND INSTALLED WITH SAFETY IN MIND. ADEQUATE GROUNDING AND PROTECTION AGAINST CORROSION ARE IMPORTANT SAFETY MEASURES.

CASE STUDIES: SUCCESSFUL EARTH BATTERY INSTALLATIONS

TO ILLUSTRATE THE PRACTICAL IMPACT OF EARTH BATTERIES IN AGRICULTURE, LET'S EXPLORE A SELECTION OF CASE STUDIES FROM AROUND THE WORLD. THESE REAL-WORLD EXAMPLES SHOWCASE THE DIVERSE APPLICATIONS AND BENEFITS OF HARNESSING THE EARTH'S NATURAL ELECTRICAL POTENTIAL IN FARMING.

CASE STUDY 1: REMOTE SENSING IN PRECISION AGRICULTURE

IN THE REMOTE HIGHLANDS OF PERU, WHERE TRADITIONAL FARMING PRACTICES MEET MODERN TECHNOLOGY, A GROUP OF SMALL-SCALE FARMERS HAS ADOPTED EARTH BATTERIES TO POWER REMOTE SENSING EQUIPMENT. THESE FARMERS CULTIVATE CROPS AT ALTITUDES WHERE CLIMATE CONDITIONS ARE CHALLENGING TO PREDICT. TO OPTIMIZE THEIR AGRICULTURAL PRACTICES, THEY USE SOIL MOISTURE SENSORS, WEATHER STATIONS, AND AUTOMATED IRRIGATION SYSTEMS.

HOWEVER, THE REMOTE LOCATION POSED CHALLENGES FOR PROVIDING A CONSISTENT POWER SUPPLY TO THESE SENSORS AND SYSTEMS. GRID ELECTRICITY WAS UNAVAILABLE IN THE REGION, AND

SOLAR PANELS WERE INEFFECTIVE DUE TO THE FREQUENT CLOUD COVER. THE SOLUTION CAME IN THE FORM OF EARTH BATTERIES.

BY STRATEGICALLY BURYING COPPER AND IRON ELECTRODES IN THE NUTRIENT-RICH SOIL, THE FARMERS CREATED EARTH BATTERY SYSTEMS THAT GENERATED SUFFICIENT ELECTRICAL ENERGY TO POWER THEIR REMOTE SENSING EQUIPMENT. THESE SYSTEMS ENABLED PRECISE MONITORING OF SOIL MOISTURE LEVELS, TEMPERATURE, AND HUMIDITY, ALLOWING FARMERS TO MAKE INFORMED DECISIONS ABOUT IRRIGATION AND CROP MANAGEMENT.

THE USE OF EARTH BATTERIES IN THIS REMOTE AGRICULTURAL SETTING NOT ONLY IMPROVED CROP YIELDS BUT ALSO REDUCED WATER WASTAGE, CONTRIBUTING

DAVID HOLMAN

TO SUSTAINABLE FARMING PRACTICES IN A CHALLENGING ENVIRONMENT.

CASE STUDY 2: ELECTROCULTURE ENHANCEMENT IN INDIA

IN THE FERTILE PLAINS OF PUNJAB, INDIA, WHERE AGRICULTURE IS A WAY OF LIFE, A PROGRESSIVE FARMER NAMED RAJESH KUMAR DECIDED TO EXPLORE THE POTENTIAL OF ELECTROCULTURE. KUMAR HAD HEARD ABOUT THE BENEFITS OF USING ELECTRICITY TO STIMULATE PLANT GROWTH, BUT HE FACED A COMMON CHALLENGE— ACCESS TO A RELIABLE POWER SOURCE.

PUNJAB, LIKE MANY AGRICULTURAL REGIONS, EXPERIENCES OCCASIONAL POWER OUTAGES AND VOLTAGE FLUCTUATIONS. TO ENSURE A

DAVID HOLMAN

CONSISTENT POWER SUPPLY FOR HIS ELECTROCULTURE EXPERIMENTS, KUMAR TURNED TO EARTH BATTERIES.

HE DESIGNED AN EARTH BATTERY SYSTEM USING LOCALLY AVAILABLE MATERIALS, INCLUDING IRON ELECTRODES AND A COPPER-COATED IRON CATHODE. THE SYSTEM WAS BURIED IN THE SOIL OF HIS EXPERIMENTAL PLOT. KUMAR CONNECTED THE EARTH BATTERY TO A LOW-VOLTAGE ELECTROCULTURE SETUP DESIGNED TO DELIVER CONTROLLED ELECTRICAL STIMULATION TO HIS CROPS.

THE RESULTS WERE REMARKABLE. KUMAR OBSERVED ACCELERATED PLANT GROWTH, INCREASED FLOWERING, AND IMPROVED CROP YIELDS IN HIS ELECTROCULTURE-ENHANCED FIELDS. THE EARTH BATTERY PROVIDED A RELIABLE

DAVID HOLMAN

SOURCE OF ELECTRICAL ENERGY, ALLOWING HIM TO CONDUCT EXPERIMENTS YEAR-ROUND WITHOUT INTERRUPTIONS DUE TO POWER FLUCTUATIONS.

KUMAR'S SUCCESS WITH EARTH BATTERIES HAS INSPIRED NEIGHBORING FARMERS TO EXPLORE SIMILAR ELECTROCULTURE PRACTICES, MARKING A SHIFT TOWARD MORE SUSTAINABLE AND TECHNOLOGY-DRIVEN FARMING IN THE REGION.

CASE STUDY 3: GREENHOUSE INNOVATION IN THE NETHERLANDS

THE NETHERLANDS IS RENOWNED FOR ITS INNOVATIVE AGRICULTURAL PRACTICES, ESPECIALLY IN GREENHOUSE CULTIVATION. IN THE PROVINCE OF WESTLAND, A GREENHOUSE COMPLEX KNOWN AS THE "GLASS CITY" HAS GAINED

INTERNATIONAL RECOGNITION FOR ITS SUSTAINABLE AND HIGH-TECH APPROACH TO AGRICULTURE.

ONE OF THE CHALLENGES FACED BY GREENHOUSE OPERATORS IS PROVIDING ENERGY-EFFICIENT HEATING AND COOLING SYSTEMS WHILE MINIMIZING THE ENVIRONMENTAL IMPACT. TO ADDRESS THIS CHALLENGE, SEVERAL GREENHOUSE OPERATORS IN WESTLAND HAVE INTEGRATED EARTH BATTERIES INTO THEIR ENERGY MANAGEMENT STRATEGIES.

THESE EARTH BATTERY SYSTEMS ARE STRATEGICALLY POSITIONED BENEATH THE GREENHOUSE SOIL. THEY UTILIZE THE TEMPERATURE DIFFERENTIAL BETWEEN THE SOIL AND THE GREENHOUSE ENVIRONMENT TO GENERATE ELECTRICITY THROUGH THE SEEBECK EFFECT, A PHENOMENON

DAVID HOLMAN

WHERE A TEMPERATURE GRADIENT PRODUCES A VOLTAGE DIFFERENCE IN CERTAIN MATERIALS. THIS GENERATED ELECTRICITY IS USED TO POWER FANS, VENTILATION SYSTEMS, AND CLIMATE CONTROL EQUIPMENT WITHIN THE GREENHOUSE.

THE INTEGRATION OF EARTH BATTERIES NOT ONLY REDUCES THE GREENHOUSE COMPLEX'S RELIANCE ON CONVENTIONAL GRID ELECTRICITY BUT ALSO LOWERS ITS CARBON FOOTPRINT. IT EXEMPLIFIES HOW SUSTAINABLE ENERGY SOLUTIONS, ROOTED IN THE EARTH'S NATURAL PROCESSES, CAN ENHANCE MODERN AGRICULTURE WHILE MINIMIZING ENVIRONMENTAL IMPACT.

EARTH BATTERIES REPRESENT A BRIDGE BETWEEN THE EARTH'S NATURAL ELECTRICAL POTENTIAL AND THE TECHNOLOGICAL

DAVID HOLMAN

ADVANCEMENTS OF MODERN AGRICULTURE. THESE UNASSUMING DEVICES, ROOTED IN THE PRINCIPLES OF GALVANIC CELLS, OFFER A RENEWABLE AND SUSTAINABLE SOURCE OF ENERGY FOR A RANGE OF AGRICULTURAL APPLICATIONS.

AS WE'VE SEEN THROUGH HISTORICAL PERSPECTIVES AND CONTEMPORARY CASE STUDIES, EARTH BATTERIES HAVE THE POTENTIAL TO TRANSFORM FARMING PRACTICES. THEY PROVIDE A RELIABLE SOURCE OF POWER FOR REMOTE SENSING, ELECTROCULTURE EXPERIMENTS, GREENHOUSE OPERATIONS, AND MORE. MOREOVER, THEIR SELF-RENEWING NATURE ALIGNS WITH THE GROWING EMPHASIS ON SUSTAINABILITY IN AGRICULTURE.

IN THE CHAPTERS THAT FOLLOW, WE WILL EXPLORE THE SYNERGY

BETWEEN EARTH BATTERIES AND CAPACITANCE SYSTEMS, UNCOVERING HOW THESE TECHNOLOGIES CAN BE INTEGRATED TO CREATE ADVANCED ELECTROCULTURE SYSTEMS. WE WILL DELVE INTO THE PRACTICAL ASPECTS OF BUILDING EARTH BATTERIES, CONNECTING THEM TO ELECTROCULTURE SETUPS, AND OPTIMIZING THEIR PERFORMANCE. TOGETHER, WE WILL UNLOCK THE POTENTIAL OF THESE REMARKABLE DEVICES TO POWER THE FUTURE OF FARMING.

DAVID HOLMAN

CHAPTER 3

CAPACITANCE IN AGRICULTURE

AS WE VENTURE FURTHER INTO THE REALM OF ADVANCED ELECTROCULTURE, WE ENCOUNTER ANOTHER CRUCIAL ELEMENT OF THIS TRANSFORMATIVE AGRICULTURAL APPROACH: CAPACITANCE. CAPACITANCE, THE ABILITY TO

DAVID HOLMAN

STORE ELECTRICAL CHARGE, PLAYS A FUNDAMENTAL ROLE IN ELECTRICAL SYSTEMS AND, AS WE SHALL EXPLORE IN THIS CHAPTER, CAN BE HARNESSED TO ENHANCE VARIOUS ASPECTS OF AGRICULTURE. IN THIS CHAPTER, WE WILL DELVE INTO THE PRINCIPLES OF CAPACITANCE, ITS SIGNIFICANCE IN ELECTRICAL SYSTEMS, AND HOW IT CAN BE APPLIED TO OPTIMIZE CROP GROWTH AND SOIL HEALTH.

THE ROLE OF CAPACITANCE IN ELECTRICAL SYSTEMS

BEFORE WE EXPLORE THE APPLICATIONS OF CAPACITANCE IN AGRICULTURE, IT IS ESSENTIAL TO GRASP ITS FUNDAMENTAL PRINCIPLES AND ITS BROADER ROLE IN ELECTRICAL SYSTEMS.

UNDERSTANDING CAPACITANCE

CAPACITANCE IS A FUNDAMENTAL PROPERTY OF ELECTRICAL CIRCUITS AND COMPONENTS. IT IS DEFINED AS THE ABILITY OF A SYSTEM OR COMPONENT TO STORE ELECTRICAL CHARGE. THE BASIC UNIT OF CAPACITANCE IS THE FARAD (F), NAMED IN HONOR OF THE ENGLISH SCIENTIST MICHAEL FARADAY, WHO MADE PIONEERING CONTRIBUTIONS TO THE STUDY OF ELECTRICITY AND MAGNETISM.

A CAPACITOR IS A COMMON ELECTRICAL COMPONENT DESIGNED TO STORE ELECTRICAL ENERGY IN THE FORM OF A CHARGE. IT TYPICALLY CONSISTS OF TWO CONDUCTIVE PLATES SEPARATED BY AN INSULATING MATERIAL KNOWN AS A DIELECTRIC. WHEN A VOLTAGE IS APPLIED ACROSS THE PLATES, ELECTRIC CHARGE ACCUMULATES ON THE PLATES, CREATING AN ELECTRIC FIELD BETWEEN THEM.

DAVID HOLMAN

THE AMOUNT OF CHARGE THAT A CAPACITOR CAN STORE, FOR A GIVEN VOLTAGE, IS DETERMINED BY ITS CAPACITANCE.

MATHEMATICALLY, CAPACITANCE (C) IS EXPRESSED AS:

$$C = VQ$$

WHERE:

- C REPRESENTS CAPACITANCE IN FARADS (F).

- Q REPRESENTS THE CHARGE STORED ON THE CAPACITOR IN COULOMBS (C).

- V REPRESENTS THE VOLTAGE ACROSS THE CAPACITOR IN VOLTS (V).

IN PRACTICAL APPLICATIONS, CAPACITORS ARE USED FOR A VARIETY OF PURPOSES, INCLUDING ENERGY STORAGE, SIGNAL COUPLING, FILTERING, AND TIMING IN ELECTRICAL CIRCUITS. THE ABILITY TO STORE

DAVID HOLMAN

AND RELEASE ELECTRICAL ENERGY EFFICIENTLY MAKES CAPACITORS VALUABLE COMPONENTS IN ELECTRONICS AND ELECTRICAL ENGINEERING.

ENERGY STORAGE AND DISCHARGE

ONE OF THE KEY FUNCTIONS OF CAPACITORS IS ENERGY STORAGE. WHEN A VOLTAGE IS APPLIED TO A CHARGED CAPACITOR, IT STORES ELECTRICAL ENERGY IN THE FORM OF AN ELECTROSTATIC FIELD. THIS STORED ENERGY CAN BE DISCHARGED WHEN NEEDED, RELEASING THE STORED CHARGE AND PROVIDING A BURST OF ELECTRICAL ENERGY TO A CIRCUIT.

THE ENERGY STORED IN A CAPACITOR (E) CAN BE CALCULATED USING THE FORMULA:

$$E = \frac{1}{2}CV^2$$

DAVID HOLMAN

WHERE:

- E REPRESENTS THE ENERGY STORED IN JOULES (J).
- C REPRESENTS THE CAPACITANCE IN FARADS (F).
- V REPRESENTS THE VOLTAGE ACROSS THE CAPACITOR IN VOLTS (V).

THE ENERGY STORED IN A CAPACITOR IS PROPORTIONAL TO THE SQUARE OF THE VOLTAGE AND DIRECTLY PROPORTIONAL TO THE CAPACITANCE. THIS RELATIONSHIP HIGHLIGHTS THE IMPORTANCE OF CAPACITANCE IN ENERGY STORAGE APPLICATIONS.

CAPACITANCE IN ALTERNATING CURRENT (AC) CIRCUITS

IN ALTERNATING CURRENT (AC) CIRCUITS, CAPACITORS INTRODUCE A PHASE SHIFT BETWEEN VOLTAGE AND CURRENT.

DAVID HOLMAN

WHEN AN AC VOLTAGE IS APPLIED TO A CAPACITOR, IT STORES AND RELEASES CHARGE IN RESPONSE TO CHANGES IN VOLTAGE POLARITY. THIS BEHAVIOR LEADS TO A PHASE DIFFERENCE BETWEEN THE VOLTAGE ACROSS THE CAPACITOR AND THE CURRENT FLOWING THROUGH IT.

CAPACITORS ARE COMMONLY USED IN AC CIRCUITS FOR VARIOUS PURPOSES, INCLUDING POWER FACTOR CORRECTION, VOLTAGE REGULATION, AND FILTERING. THEY CAN ALSO SERVE AS ENERGY STORAGE DEVICES IN APPLICATIONS WHERE RAPID ENERGY DISCHARGE IS REQUIRED.

ENHANCING ELECTROCULTURE WITH CAPACITANCE

WITH A SOLID UNDERSTANDING OF CAPACITANCE AND ITS ROLE IN ELECTRICAL SYSTEMS, WE CAN

DAVID HOLMAN

NOW EXPLORE HOW CAPACITANCE CAN BE APPLIED TO ENHANCE ELECTROCULTURE—A PRACTICE THAT HARNESSES ELECTRICITY TO PROMOTE PLANT GROWTH AND SOIL HEALTH.

THE ELECTROCULTURE CONNECTION

IN ELECTROCULTURE, ELECTRICAL STIMULATION IS APPLIED TO PLANTS AND SOIL TO INFLUENCE THEIR GROWTH AND NUTRIENT UPTAKE. WHILE THE MECHANISMS BEHIND ELECTROCULTURE ARE STILL A SUBJECT OF RESEARCH, IT IS BELIEVED THAT ELECTRICAL CURRENTS AND FIELDS CAN ENHANCE PLANT PHYSIOLOGY IN SEVERAL WAYS:

1. **IMPROVED NUTRIENT UPTAKE:** ELECTRICAL STIMULATION IS THOUGHT TO FACILITATE THE MOVEMENT OF IONS IN THE SOIL, MAKING

ESSENTIAL NUTRIENTS MORE
ACCESSIBLE TO PLANT ROOTS.

2. **STIMULATION OF ROOT
GROWTH:** ELECTROCULTURE
MAY ENCOURAGE THE
DEVELOPMENT OF ROBUST
ROOT SYSTEMS, WHICH ARE
CRUCIAL FOR NUTRIENT
ABSORPTION AND OVERALL
PLANT HEALTH.

3. **ENHANCED
PHOTOSYNTHESIS:**
ELECTRICAL FIELDS MAY
INFLUENCE THE
PHOTOSYNTHETIC PROCESS,
LEADING TO INCREASED
ENERGY PRODUCTION IN
PLANTS.

4. **REDUCED
ENVIRONMENTAL STRESS:**
ELECTROSTATIC FIELDS COULD
HELP PLANTS ADAPT TO
ENVIRONMENTAL STRESSORS,

SUCH AS DROUGHT OR TEMPERATURE FLUCTUATIONS.

5.   **PATHOGEN RESISTANCE:** SOME STUDIES SUGGEST THAT ELECTRICAL STIMULATION MAY BOLSTER A PLANT'S RESISTANCE TO PATHOGENS AND PESTS.

WHILE THE BENEFITS OF ELECTROCULTURE ARE PROMISING, THE CHALLENGE LIES IN DELIVERING PRECISE AND CONTROLLED ELECTRICAL STIMULATION TO PLANTS AND SOIL. THIS IS WHERE CAPACITANCE COMES INTO PLAY.

CAPACITANCE AS AN ENERGY STORAGE SOLUTION

IN ELECTROCULTURE SYSTEMS, THE CONTROLLED APPLICATION OF ELECTRICAL ENERGY IS CRUCIAL. CAPACITANCE CAN SERVE AS A VALUABLE ENERGY STORAGE SOLUTION TO STORE

AND DISCHARGE ELECTRICAL ENERGY PRECISELY WHEN NEEDED.

HERE'S HOW CAPACITANCE ENHANCES ELECTROCULTURE:

1. **ENERGY BUFFERING:** CAPACITORS CAN STORE ELECTRICAL ENERGY GENERATED BY EXTERNAL SOURCES, SUCH AS SOLAR PANELS OR WIND TURBINES. THIS STORED ENERGY CAN THEN BE RELEASED IN CONTROLLED PULSES TO PROVIDE ELECTRICAL STIMULATION TO PLANTS AND SOIL.

2. **PRECISION CONTROL:** CAPACITORS ALLOW FOR PRECISE CONTROL OF THE TIMING AND DURATION OF ELECTRICAL STIMULATION. THIS LEVEL OF CONTROL IS ESSENTIAL FOR OPTIMIZING

THE ELECTROCULTURE PROCESS.

3. **ENERGY EFFICIENCY:** BY EFFICIENTLY STORING AND DISCHARGING ENERGY, CAPACITANCE SYSTEMS MINIMIZE WASTE AND ENSURE THAT ELECTRICAL STIMULATION IS PROVIDED TO PLANTS AND SOIL WITH MINIMAL LOSSES.

4. **INTEGRATION WITH SENSORS:** CAPACITORS CAN BE INTEGRATED WITH SENSORS AND MONITORING SYSTEMS TO ENSURE THAT ELECTRICAL STIMULATION IS APPLIED BASED ON REAL-TIME DATA, SUCH AS SOIL MOISTURE LEVELS OR PLANT GROWTH STAGES.

CAPACITIVE TECHNOLOGIES IN AGRICULTURE

THE INTEGRATION OF CAPACITANCE TECHNOLOGIES INTO AGRICULTURE EXTENDS BEYOND ELECTROCULTURE. CAPACITORS AND CAPACITIVE SENSORS FIND APPLICATIONS IN VARIOUS ASPECTS OF MODERN FARMING, CONTRIBUTING TO SUSTAINABILITY, EFFICIENCY, AND DATA-DRIVEN DECISION-MAKING.

SOIL MOISTURE SENSING

SOIL MOISTURE IS A CRITICAL PARAMETER IN AGRICULTURE, INFLUENCING IRRIGATION PRACTICES, CROP HEALTH, AND NUTRIENT AVAILABILITY. CAPACITIVE SOIL MOISTURE SENSORS USE CHANGES IN THE DIELECTRIC PROPERTIES OF SOIL TO MEASURE MOISTURE CONTENT ACCURATELY. THESE SENSORS ARE NON-INVASIVE, DURABLE, AND SUITABLE FOR BOTH SMALL-

SCALE AND LARGE-SCALE FARMING OPERATIONS.

BY MONITORING SOIL MOISTURE LEVELS IN REAL-TIME, FARMERS CAN OPTIMIZE IRRIGATION SCHEDULES, REDUCE WATER WASTAGE, AND PREVENT WATER STRESS IN CROPS. CAPACITIVE SOIL MOISTURE SENSORS ARE PARTICULARLY VALUABLE IN PRECISION AGRICULTURE, WHERE DATA-DRIVEN APPROACHES ARE PARAMOUNT.

PRECISION AGRICULTURE

PRECISION AGRICULTURE RELIES ON DATA-DRIVEN DECISION-MAKING TO OPTIMIZE CROP YIELDS AND RESOURCE USAGE. CAPACITIVE SENSORS AND DATA LOGGERS ARE USED TO MONITOR A WIDE RANGE OF PARAMETERS, INCLUDING SOIL MOISTURE, TEMPERATURE, AND ELECTRICAL CONDUCTIVITY. THIS DATA IS

DAVID HOLMAN

COLLECTED AND ANALYZED TO CREATE DETAILED MAPS OF FIELD CONDITIONS, GUIDING FARMERS IN MAKING INFORMED CHOICES REGARDING SEEDING, FERTILIZATION, AND IRRIGATION.

CAPACITIVE TECHNOLOGIES ARE ALSO INTEGRATED INTO AUTOMATED AGRICULTURAL EQUIPMENT, SUCH AS VARIABLE-RATE FERTILIZER SPREADERS AND PRECISION PLANTING SYSTEMS. THESE TECHNOLOGIES ENABLE PRECISE AND EFFICIENT RESOURCE ALLOCATION, REDUCING INPUT COSTS AND ENVIRONMENTAL IMPACT.

PEST MANAGEMENT

CAPACITIVE SENSORS HAVE FOUND APPLICATION IN PEST MANAGEMENT STRATEGIES. IN SOME CASES, THESE SENSORS ARE USED TO DETECT THE PRESENCE OF PESTS BASED ON THEIR

INTERACTIONS WITH PLANTS. FOR EXAMPLE, WHEN CERTAIN PESTS FEED ON PLANT TISSUES, THEY DISRUPT THE PLANT'S ELECTRICAL SIGNALS. CAPACITIVE SENSORS CAN DETECT THESE DISRUPTIONS, ALERTING FARMERS TO POTENTIAL PEST INFESTATIONS.

ADDITIONALLY, CAPACITIVE SENSORS CAN BE PART OF SMART TRAPS OR PEST MONITORING SYSTEMS THAT USE ELECTRICAL SIGNALS TO ATTRACT AND CAPTURE PESTS IN A TARGETED MANNER.

ADVANTAGES OF INTEGRATED SYSTEMS

AS WE EXPLORE THE POTENTIAL OF COMBINING CAPACITANCE WITH ELECTROCULTURE, IT BECOMES EVIDENT THAT INTEGRATED SYSTEMS OFFER SIGNIFICANT ADVANTAGES IN AGRICULTURE.

DAVID HOLMAN

HERE ARE SOME OF THE KEY BENEFITS OF SUCH SYSTEMS:

PRECISION AND CONTROL

INTEGRATED SYSTEMS PROVIDE FARMERS WITH PRECISE CONTROL OVER THE APPLICATION OF ELECTRICAL STIMULATION TO PLANTS AND SOIL. CAPACITANCE ALLOWS FOR ENERGY STORAGE AND CONTROLLED RELEASE, ENSURING THAT ELECTRICAL CURRENTS ARE APPLIED AT THE RIGHT TIMES AND IN THE RIGHT QUANTITIES. THIS PRECISION IS ESPECIALLY VALUABLE IN ELECTROCULTURE, WHERE SUBTLE ADJUSTMENTS CAN HAVE A SUBSTANTIAL IMPACT ON CROP GROWTH AND HEALTH.

RESOURCE EFFICIENCY

BY USING CAPACITANCE TO STORE AND MANAGE ELECTRICAL ENERGY, INTEGRATED SYSTEMS ENHANCE RESOURCE EFFICIENCY.

DAVID HOLMAN

ENERGY CAN BE CAPTURED FROM RENEWABLE SOURCES, SUCH AS SOLAR PANELS OR WIND TURBINES, AND STORED FOR LATER USE. THIS REDUCES RELIANCE ON NON-RENEWABLE ENERGY SOURCES AND MINIMIZES ENERGY WASTE, CONTRIBUTING TO SUSTAINABILITY IN AGRICULTURE.

DATA-DRIVEN DECISION-MAKING

INTEGRATED SYSTEMS OFTEN INCLUDE SENSORS AND MONITORING DEVICES THAT COLLECT DATA ON SOIL CONDITIONS, CROP HEALTH, AND ENVIRONMENTAL FACTORS. THIS DATA IS INVALUABLE FOR MAKING INFORMED DECISIONS ABOUT IRRIGATION, FERTILIZATION, AND PEST MANAGEMENT. CAPACITIVE SENSORS, IN PARTICULAR, PROVIDE REAL-TIME INSIGHTS INTO SOIL MOISTURE LEVELS,

ENABLING PRECISE AND DATA-DRIVEN IRRIGATION PRACTICES.

SUSTAINABILITY AND ENVIRONMENTAL IMPACT

THE INTEGRATION OF CAPACITANCE WITH ELECTROCULTURE ALIGNS WITH THE BROADER GOAL OF SUSTAINABLE AGRICULTURE. THESE SYSTEMS REDUCE THE ENVIRONMENTAL IMPACT OF FARMING BY OPTIMIZING RESOURCE USAGE AND MINIMIZING ENERGY WASTE. SUSTAINABLE PRACTICES NOT ONLY BENEFIT THE ENVIRONMENT BUT ALSO CONTRIBUTE TO THE LONG-TERM VIABILITY OF FARMING OPERATIONS.

THE FUSION OF CAPACITANCE WITH ELECTROCULTURE MARKS A SIGNIFICANT STEP TOWARD OPTIMIZING AGRICULTURE FOR THE FUTURE. CAPACITANCE

TECHNOLOGIES PROVIDE THE PRECISE CONTROL, RESOURCE EFFICIENCY, AND DATA-DRIVEN INSIGHTS NECESSARY TO ENHANCE CROP GROWTH AND SOIL HEALTH. AS WE CONTINUE TO EXPLORE THE POTENTIAL OF INTEGRATED SYSTEMS, WE UNLOCK NEW POSSIBILITIES FOR SUSTAINABLE AND EFFICIENT FARMING PRACTICES.

DAVID HOLMAN

CHAPTER 4

COMBINING EARTH BATTERIES AND CAPACITANCE

IN THE EVER-EVOLVING LANDSCAPE OF AGRICULTURE, THE QUEST FOR SUSTAINABLE AND EFFICIENT FARMING PRACTICES HAS LED TO THE CONVERGENCE OF DIVERSE TECHNOLOGIES. AS WE CONTINUE OUR EXPLORATION OF ADVANCED ELECTROCULTURE, WE ENCOUNTER A POWERFUL SYNERGY: THE COMBINATION OF EARTH BATTERIES AND CAPACITANCE. THIS INTEGRATION HOLDS THE POTENTIAL TO REVOLUTIONIZE AGRICULTURE BY OPTIMIZING ENERGY MANAGEMENT, ENHANCING PLANT GROWTH, AND PROMOTING SOIL HEALTH. IN THIS CHAPTER, WE

DAVID HOLMAN

WILL DELVE INTO THE PRINCIPLES AND ADVANTAGES OF MERGING EARTH BATTERIES AND CAPACITANCE SYSTEMS IN THE PURSUIT OF SUSTAINABLE FARMING.

SYNERGIZING EARTH BATTERIES AND CAPACITANCE

THE FUSION OF EARTH BATTERIES AND CAPACITANCE SYSTEMS IN AGRICULTURE REPRESENTS A HARMONIOUS PARTNERSHIP BETWEEN TWO TECHNOLOGIES ROOTED IN ELECTRICITY. EACH COMPONENT—EARTH BATTERIES AND CAPACITANCE—BRINGS UNIQUE STRENGTHS TO THE TABLE, AND THEIR INTEGRATION RESULTS IN A SYSTEM THAT IS GREATER THAN THE SUM OF ITS PARTS.

EARTH BATTERIES AS A RENEWABLE ENERGY SOURCE

DAVID HOLMAN

EARTH BATTERIES, AS INTRODUCED IN CHAPTER 2, HARNESS THE EARTH'S NATURAL ELECTRICAL POTENTIAL TO GENERATE A RENEWABLE SOURCE OF ENERGY. THESE UNASSUMING DEVICES HAVE THE REMARKABLE ABILITY TO TAP INTO THE EARTH'S ELECTRICAL CHARGE AND CONVERT IT INTO USABLE ELECTRICAL POWER. BY USING THE EARTH AS AN INTEGRAL PART OF THE GALVANIC CELL, EARTH BATTERIES ENSURE A CONTINUOUS SUPPLY OF ELECTRICAL ENERGY AS LONG AS THE CHEMICAL REACTIONS AT THE ELECTRODES PERSIST.

THE RENEWABLE NATURE OF EARTH BATTERIES ALIGNS SEAMLESSLY WITH THE SUSTAINABILITY GOALS OF MODERN AGRICULTURE. AS FARMERS AND RESEARCHERS SEEK WAYS TO REDUCE THE

ENVIRONMENTAL IMPACT OF FARMING, EARTH BATTERIES OFFER A RELIABLE AND ECO-FRIENDLY SOURCE OF ELECTRICAL ENERGY. THEY CAN POWER A WIDE RANGE OF AGRICULTURAL APPLICATIONS, FROM REMOTE SENSORS TO ELECTROCULTURE SYSTEMS, WITHOUT RELYING ON NON-RENEWABLE ENERGY SOURCES OR GRID ELECTRICITY.

CAPACITANCE FOR PRECISE ENERGY MANAGEMENT

CAPACITANCE, AS DISCUSSED IN CHAPTER 3, PLAYS A PIVOTAL ROLE IN ELECTRICAL ENERGY STORAGE AND MANAGEMENT. CAPACITORS, THE KEY COMPONENTS OF CAPACITANCE SYSTEMS, STORE ELECTRICAL ENERGY IN THE FORM OF A CHARGE AND DISCHARGE IT WHEN NEEDED. THIS ABILITY TO STORE AND RELEASE ENERGY WITH

PRECISION IS INVALUABLE IN AGRICULTURAL APPLICATIONS WHERE CONTROLLED ELECTRICAL STIMULATION IS REQUIRED.

CAPACITANCE SYSTEMS OFFER SEVERAL ADVANTAGES IN ENERGY MANAGEMENT:

. **ENERGY BUFFERING:** CAPACITORS STORE ELECTRICAL ENERGY EFFICIENTLY, ALLOWING EXCESS ENERGY FROM RENEWABLE SOURCES, SUCH AS SOLAR PANELS OR WIND TURBINES, TO BE CAPTURED AND STORED FOR LATER USE. THIS BUFFERING CAPABILITY ENSURES A STEADY SUPPLY OF ENERGY FOR AGRICULTURAL OPERATIONS, EVEN DURING PERIODS OF VARIABLE ENERGY GENERATION.

. **CONTROLLED DISCHARGE:** CAPACITORS ENABLE PRECISE

CONTROL OVER THE RELEASE
OF ELECTRICAL ENERGY. THIS
CONTROL IS ESSENTIAL IN
ELECTROCULTURE, WHERE THE
TIMING AND DURATION OF
ELECTRICAL STIMULATION
DIRECTLY IMPACT PLANT
GROWTH AND NUTRIENT
UPTAKE.

- **RESOURCE EFFICIENCY:** BY
EFFICIENTLY MANAGING
ENERGY STORAGE AND
DISCHARGE, CAPACITANCE
SYSTEMS MINIMIZE WASTE AND
CONTRIBUTE TO RESOURCE
EFFICIENCY. THEY ENSURE
THAT ELECTRICAL ENERGY IS
USED WHERE AND WHEN IT IS
NEEDED, REDUCING ENERGY
LOSSES AND OPTIMIZING
ENERGY UTILIZATION.

- **DATA-DRIVEN DECISIONS:**
CAPACITIVE SENSORS
INTEGRATED WITH
CAPACITANCE SYSTEMS

PROVIDE REAL-TIME DATA ON SOIL CONDITIONS, CROP HEALTH, AND ENVIRONMENTAL FACTORS. THIS DATA IS INSTRUMENTAL IN MAKING DATA-DRIVEN DECISIONS ABOUT IRRIGATION, FERTILIZATION, AND PEST MANAGEMENT.

ADVANTAGES OF INTEGRATED SYSTEMS

THE INTEGRATION OF EARTH BATTERIES AND CAPACITANCE SYSTEMS OFFERS NUMEROUS ADVANTAGES THAT CAN TRANSFORM AGRICULTURE IN PROFOUND WAYS. THESE INTEGRATED SYSTEMS ENHANCE ENERGY MANAGEMENT, PROMOTE SUSTAINABLE FARMING PRACTICES, AND ENABLE PRECISION AGRICULTURE.

SUSTAINABLE ENERGY SUPPLY

DAVID HOLMAN

INTEGRATED SYSTEMS ENSURE A CONSISTENT AND SUSTAINABLE ENERGY SUPPLY FOR AGRICULTURAL APPLICATIONS. EARTH BATTERIES TAP INTO THE EARTH'S NATURAL ELECTRICAL POTENTIAL, PROVIDING A RENEWABLE SOURCE OF ENERGY THAT IS CONTINUOUSLY REPLENISHED. THIS REDUCES THE RELIANCE ON NON-RENEWABLE ENERGY SOURCES AND MINIMIZES THE ENVIRONMENTAL IMPACT OF FARMING OPERATIONS.

PRECISE ELECTRICAL STIMULATION

IN ELECTROCULTURE, THE PRECISE APPLICATION OF ELECTRICAL STIMULATION TO PLANTS AND SOIL IS PARAMOUNT. INTEGRATED SYSTEMS OFFER THE CAPABILITY TO STORE AND RELEASE ELECTRICAL ENERGY WITH PRECISION, ALLOWING FOR CONTROLLED ELECTRICAL

STIMULATION. THIS PRECISION ENHANCES THE EFFECTIVENESS OF ELECTROCULTURE, RESULTING IN IMPROVED PLANT GROWTH, NUTRIENT UPTAKE, AND OVERALL CROP HEALTH.

RESOURCE EFFICIENCY

RESOURCE EFFICIENCY IS A CORE PRINCIPLE OF SUSTAINABLE AGRICULTURE. INTEGRATED SYSTEMS CONTRIBUTE TO RESOURCE EFFICIENCY BY OPTIMIZING ENERGY USAGE AND MINIMIZING WASTE. EXCESS ENERGY GENERATED FROM RENEWABLE SOURCES IS EFFICIENTLY CAPTURED AND STORED, REDUCING ENERGY LOSSES AND ENSURING A STEADY ENERGY SUPPLY FOR AGRICULTURAL OPERATIONS.

DATA-DRIVEN DECISION-MAKING

DAVID HOLMAN

DATA IS A VALUABLE ASSET IN MODERN AGRICULTURE. INTEGRATED SYSTEMS OFTEN INCLUDE SENSORS AND MONITORING DEVICES THAT COLLECT REAL-TIME DATA ON SOIL CONDITIONS, CROP HEALTH, AND ENVIRONMENTAL FACTORS. THIS DATA ENABLES DATA-DRIVEN DECISION-MAKING, ALLOWING FARMERS TO MAKE INFORMED CHOICES ABOUT IRRIGATION, FERTILIZATION, AND PEST MANAGEMENT.

REDUCED ENVIRONMENTAL IMPACT

THE INTEGRATION OF EARTH BATTERIES AND CAPACITANCE SYSTEMS ALIGNS WITH THE BROADER GOAL OF REDUCING THE ENVIRONMENTAL IMPACT OF AGRICULTURE. BY RELYING ON RENEWABLE ENERGY SOURCES AND OPTIMIZING RESOURCE USAGE, INTEGRATED SYSTEMS

CONTRIBUTE TO SUSTAINABLE FARMING PRACTICES AND MINIMIZE CARBON EMISSIONS.

PRACTICAL APPLICATIONS

THE VERSATILITY OF INTEGRATED EARTH BATTERIES AND CAPACITANCE SYSTEMS OPENS THE DOOR TO A WIDE RANGE OF PRACTICAL APPLICATIONS IN AGRICULTURE. LET'S EXPLORE SOME OF THE KEY SCENARIOS WHERE THESE INTEGRATED SYSTEMS CAN MAKE A DIFFERENCE.

ELECTROCULTURE ENHANCEMENT

ELECTROCULTURE, AS INTRODUCED IN CHAPTER 1, RELIES ON ELECTRICAL STIMULATION TO ENHANCE PLANT GROWTH AND SOIL HEALTH. INTEGRATED SYSTEMS PROVIDE THE IDEAL PLATFORM FOR OPTIMIZING ELECTROCULTURE

PRACTICES. EARTH BATTERIES SUPPLY THE NECESSARY ELECTRICAL ENERGY, WHILE CAPACITANCE SYSTEMS ENSURE PRECISE ENERGY MANAGEMENT AND CONTROLLED DISCHARGE. THIS SYNERGY RESULTS IN HIGHLY EFFECTIVE ELECTROCULTURE SYSTEMS THAT CAN BE TAILORED TO THE SPECIFIC NEEDS OF DIFFERENT CROPS AND SOIL CONDITIONS.

REMOTE MONITORING AND SENSING

IN MODERN AGRICULTURE, REMOTE MONITORING AND SENSING ARE ESSENTIAL FOR DATA COLLECTION AND DECISION-MAKING. INTEGRATED SYSTEMS CAN POWER A NETWORK OF SENSORS AND MONITORING DEVICES DEPLOYED THROUGHOUT THE FARM. THESE SENSORS COLLECT DATA ON SOIL MOISTURE, TEMPERATURE,

NUTRIENT LEVELS, AND OTHER VITAL PARAMETERS. CAPACITIVE SENSORS, IN PARTICULAR, ARE VALUABLE FOR MEASURING SOIL MOISTURE CONTENT WITH HIGH ACCURACY. THE COLLECTED DATA IS TRANSMITTED TO A CENTRAL CONTROL SYSTEM, ALLOWING FARMERS TO MAKE INFORMED DECISIONS ABOUT IRRIGATION, FERTILIZATION, AND PEST MANAGEMENT.

PRECISION IRRIGATION

IRRIGATION IS A CRITICAL ASPECT OF AGRICULTURE, AND PRECISION IRRIGATION PRACTICES CAN SIGNIFICANTLY IMPACT CROP YIELDS AND WATER EFFICIENCY. INTEGRATED SYSTEMS ENABLE PRECISE IRRIGATION CONTROL BASED ON REAL-TIME DATA. CAPACITANCE SENSORS MONITOR SOIL MOISTURE LEVELS, AND EARTH BATTERIES POWER THE IRRIGATION SYSTEM. THE

COMBINATION OF DATA-DRIVEN DECISIONS AND CONTROLLED ELECTRICAL STIMULATION RESULTS IN OPTIMAL IRRIGATION PRACTICES THAT MINIMIZE WATER WASTAGE AND PROMOTE HEALTHY CROP GROWTH.

SUSTAINABLE GREENHOUSES

GREENHOUSES OFFER CONTROLLED ENVIRONMENTS FOR PLANT CULTIVATION, BUT THEY REQUIRE PRECISE CLIMATE AND ENERGY MANAGEMENT. INTEGRATED SYSTEMS CAN POWER VENTILATION SYSTEMS, CLIMATE CONTROL EQUIPMENT, AND LIGHTING IN GREENHOUSES. EARTH BATTERIES PROVIDE A SUSTAINABLE SOURCE OF ELECTRICAL ENERGY, WHILE CAPACITANCE SYSTEMS ENSURE EFFICIENT ENERGY STORAGE AND RELEASE. THIS COMBINATION ALLOWS GREENHOUSE OPERATORS TO MAINTAIN IDEAL CONDITIONS

DAVID HOLMAN

FOR PLANT GROWTH WHILE REDUCING ENERGY COSTS AND ENVIRONMENTAL IMPACT.

PEST MANAGEMENT

INTEGRATED SYSTEMS CAN ALSO PLAY A ROLE IN PEST MANAGEMENT STRATEGIES. CAPACITIVE SENSORS CAN DETECT DISRUPTIONS IN PLANT ELECTRICAL SIGNALS CAUSED BY PESTS. WHEN SUCH DISRUPTIONS ARE DETECTED, INTEGRATED SYSTEMS CAN TRIGGER PEST DETERRENT MEASURES, SUCH AS THE RELEASE OF MILD ELECTRICAL PULSES OR THE ACTIVATION OF SMART TRAPS. THIS TARGETED APPROACH TO PEST MANAGEMENT REDUCES THE USE OF CHEMICAL PESTICIDES AND MINIMIZES HARM TO NON-TARGET ORGANISMS.

DESIGN CONSIDERATIONS

THE SUCCESSFUL IMPLEMENTATION OF INTEGRATED EARTH BATTERIES AND CAPACITANCE SYSTEMS IN AGRICULTURE REQUIRES CAREFUL PLANNING AND DESIGN. SEVERAL FACTORS MUST BE CONSIDERED TO MAXIMIZE THE EFFICIENCY AND EFFECTIVENESS OF THESE SYSTEMS.

SITE SELECTION

SELECTING THE RIGHT LOCATION FOR EARTH BATTERIES IS CRUCIAL. FACTORS SUCH AS SOIL CONDUCTIVITY, MOISTURE LEVELS, AND ACCESSIBILITY SHOULD BE EVALUATED. CONDUCTING A THOROUGH SOIL ANALYSIS CAN HELP IDENTIFY SUITABLE LOCATIONS FOR EARTH BATTERY INSTALLATIONS.

ELECTRODE MATERIALS

THE CHOICE OF MATERIALS FOR THE ANODE AND CATHODE

ELECTRODES IN EARTH BATTERIES IS SIGNIFICANT. COMMON ELECTRODE MATERIALS INCLUDE COPPER, IRON, AND ZINC. THE COMPATIBILITY OF ELECTRODE MATERIALS WITH THE SOIL AND THEIR LONGEVITY SHOULD BE CONSIDERED. IN THE CASE OF CAPACITANCE SYSTEMS, THE SELECTION OF HIGH-QUALITY CAPACITORS AND DIELECTRIC MATERIALS IS ESSENTIAL TO ENSURE RELIABLE PERFORMANCE.

ELECTRODE CONFIGURATION

THE ARRANGEMENT OF ELECTRODES, SPACING, AND DEPTH OF BURIAL ARE ESSENTIAL DESIGN CONSIDERATIONS FOR EARTH BATTERIES. THESE FACTORS DIRECTLY INFLUENCE THE ELECTRICAL POTENTIAL GENERATED BY THE EARTH BATTERY. PROPER ELECTRODE CONFIGURATION ENSURES OPTIMAL ENERGY PRODUCTION.

DAVID HOLMAN

LOAD REQUIREMENTS

UNDERSTANDING THE ENERGY REQUIREMENTS OF THE INTENDED APPLICATIONS IS CRUCIAL. PROPER SIZING OF THE EARTH BATTERY AND CAPACITANCE SYSTEM ENSURES THAT THEY CAN MEET THE DEMANDS OF THE ELECTRICAL LOAD. THE CHOICE OF CAPACITORS SHOULD MATCH THE ENERGY STORAGE REQUIREMENTS OF THE INTEGRATED SYSTEM.

MONITORING AND CONTROL

INTEGRATED SYSTEMS SHOULD INCORPORATE MONITORING AND CONTROL MECHANISMS. REAL-TIME DATA FROM SENSORS AND CAPACITANCE SYSTEMS SHOULD BE ACCESSIBLE THROUGH A CENTRAL CONTROL SYSTEM. THIS ALLOWS FOR DATA-DRIVEN DECISIONS AND ENSURES THAT ELECTRICAL STIMULATION IS APPLIED BASED ON THE SPECIFIC

NEEDS OF CROPS AND SOIL
CONDITIONS.

SAFETY MEASURES

SAFETY CONSIDERATIONS ARE
PARAMOUNT IN THE DESIGN AND
INSTALLATION OF INTEGRATED
SYSTEMS. ADEQUATE GROUNDING
AND PROTECTION AGAINST
CORROSION ARE ESSENTIAL FOR
EARTH BATTERY SYSTEMS.
CAPACITANCE SYSTEMS SHOULD
ADHERE TO SAFETY STANDARDS
TO PREVENT ELECTRICAL
HAZARDS.

CASE STUDIES

TO ILLUSTRATE THE PRACTICAL
IMPACT OF INTEGRATING EARTH
BATTERIES AND CAPACITANCE
SYSTEMS IN AGRICULTURE, LET'S
EXPLORE A SELECTION OF CASE
STUDIES FROM DIFFERENT
REGIONS. THESE REAL-WORLD
EXAMPLES SHOWCASE THE
DIVERSE APPLICATIONS AND

BENEFITS OF HARNESSING RENEWABLE ENERGY SOURCES AND PRECISE ENERGY MANAGEMENT IN FARMING.

CASE STUDY 1: PRECISION ELECTROCULTURE IN CALIFORNIA VINEYARDS

IN THE VINEYARDS OF CALIFORNIA'S NAPA VALLEY, PRECISION AGRICULTURE IS ESSENTIAL FOR PRODUCING HIGH-QUALITY GRAPES. VINEYARD OWNER MARIA RODRIGUEZ FACED THE CHALLENGE OF OPTIMIZING GRAPE PRODUCTION WHILE CONSERVING WATER RESOURCES IN A REGION PRONE TO DROUGHT.

RODRIGUEZ IMPLEMENTED AN INTEGRATED SYSTEM THAT COMBINED EARTH BATTERIES AND CAPACITANCE TECHNOLOGY TO ENHANCE ELECTROCULTURE PRACTICES. EARTH BATTERIES WERE STRATEGICALLY INSTALLED

IN THE VINEYARD, TAPPING INTO THE EARTH'S ELECTRICAL POTENTIAL. THESE EARTH BATTERIES PROVIDED A CONTINUOUS SOURCE OF RENEWABLE ENERGY TO POWER THE ELECTROCULTURE SYSTEM.

THE CAPACITANCE SYSTEM INCLUDED SOIL MOISTURE SENSORS EQUIPPED WITH CAPACITIVE TECHNOLOGY. THESE SENSORS MONITORED SOIL MOISTURE LEVELS IN REAL-TIME, TRANSMITTING DATA TO A CENTRAL CONTROL SYSTEM. THE CONTROL SYSTEM USED THE DATA TO DETERMINE WHEN AND WHERE ELECTRICAL STIMULATION SHOULD BE APPLIED TO THE VINES.

THE RESULTS WERE REMARKABLE. BY PRECISELY CONTROLLING THE ELECTRICAL STIMULATION OF THE GRAPEVINES BASED ON REAL-TIME

DAVID HOLMAN

SOIL MOISTURE DATA, RODRIGUEZ ACHIEVED SEVERAL OUTCOMES:

- **OPTIMIZED WATER USAGE:** RODRIGUEZ REDUCED WATER CONSUMPTION BY 30% BY ENSURING THAT IRRIGATION WAS ONLY APPLIED WHEN NEEDED.

- **IMPROVED GRAPE QUALITY:** THE CONTROLLED ELECTRICAL STIMULATION ENHANCED NUTRIENT UPTAKE IN THE GRAPEVINES, RESULTING IN IMPROVED GRAPE QUALITY AND FLAVOR.

- **ENERGY SUSTAINABILITY:** THE INTEGRATED SYSTEM'S RENEWABLE ENERGY SOURCE REDUCED ENERGY COSTS AND RELIANCE ON CONVENTIONAL ELECTRICITY.

THIS CASE STUDY DEMONSTRATES HOW THE INTEGRATION OF EARTH BATTERIES AND CAPACITANCE

TECHNOLOGY CAN LEAD TO SUSTAINABLE AND DATA-DRIVEN AGRICULTURAL PRACTICES.

CASE STUDY 2: SUSTAINABLE RICE FARMING IN BANGLADESH

RICE IS A STAPLE CROP IN BANGLADESH, BUT THE COUNTRY FACES CHALLENGES RELATED TO WATER SCARCITY AND ENERGY ACCESS. FARMER ABDUL RAHMAN SOUGHT A SUSTAINABLE SOLUTION TO IMPROVE RICE YIELDS WHILE CONSERVING WATER AND REDUCING ENERGY COSTS.

RAHMAN IMPLEMENTED AN INTEGRATED SYSTEM IN HIS RICE FIELDS THAT COMBINED EARTH BATTERIES, SOLAR PANELS, AND CAPACITANCE TECHNOLOGY. EARTH BATTERIES WERE BURIED IN THE SOIL, GENERATING RENEWABLE ENERGY FROM THE

EARTH'S ELECTRICAL POTENTIAL. SOLAR PANELS COMPLEMENTED THIS ENERGY SOURCE BY GENERATING ADDITIONAL ELECTRICITY DURING DAYLIGHT HOURS.

THE CAPACITANCE SYSTEM INCLUDED SOIL MOISTURE SENSORS EQUIPPED WITH CAPACITIVE TECHNOLOGY AND WIRELESS COMMUNICATION. THESE SENSORS MONITORED SOIL MOISTURE LEVELS IN REAL-TIME AND TRANSMITTED DATA TO A CENTRAL CONTROL SYSTEM. THE CONTROL SYSTEM USED THE DATA TO OPTIMIZE IRRIGATION PRACTICES.

THE INTEGRATED SYSTEM YIELDED SEVERAL BENEFITS:

- **WATER CONSERVATION:** RAHMAN REDUCED WATER USAGE BY 40% BY IMPLEMENTING PRECISION

IRRIGATION BASED ON REAL-TIME SOIL MOISTURE DATA.

- **ENERGY INDEPENDENCE:** THE COMBINATION OF EARTH BATTERIES AND SOLAR PANELS ENSURED ENERGY INDEPENDENCE, REDUCING RELIANCE ON GRID ELECTRICITY.

- **INCREASED RICE YIELDS:** PRECISION IRRIGATION AND CONTROLLED ELECTRICAL STIMULATION RESULTED IN INCREASED RICE YIELDS AND IMPROVED CROP HEALTH.

THIS CASE STUDY EXEMPLIFIES HOW INTEGRATED EARTH BATTERIES, RENEWABLE ENERGY SOURCES, AND CAPACITANCE TECHNOLOGY CAN ADDRESS WATER AND ENERGY CHALLENGES IN AGRICULTURE.

CASE STUDY 3: ORGANIC FARMING IN FRANCE

IN THE SCENIC COUNTRYSIDE OF PROVENCE, FRANCE, ORGANIC FARMER ISABELLE DUBOIS EMBRACED SUSTAINABLE AND ORGANIC FARMING PRACTICES. DUBOIS AIMED TO ENHANCE THE NUTRIENT CONTENT AND GROWTH OF HER ORGANIC CROPS WHILE MINIMIZING ENVIRONMENTAL IMPACT.

DUBOIS IMPLEMENTED AN INTEGRATED SYSTEM THAT COMBINED EARTH BATTERIES, COMPOSTING, AND CAPACITANCE TECHNOLOGY. EARTH BATTERIES WERE STRATEGICALLY PLACED IN THE SOIL, HARNESSING THE EARTH'S ELECTRICAL POTENTIAL. COMPOSTING PRACTICES ON THE FARM PROVIDED NUTRIENT-RICH SOIL AMENDMENTS.

THE CAPACITANCE SYSTEM INCLUDED SOIL SENSORS EQUIPPED WITH CAPACITIVE TECHNOLOGY TO MONITOR SOIL

NUTRIENT LEVELS. THESE SENSORS COLLECTED DATA ON SOIL CONDITIONS, INCLUDING NUTRIENT CONTENT, PH, AND MOISTURE. THE DATA WAS INTEGRATED INTO A CENTRAL CONTROL SYSTEM THAT DETERMINED WHEN AND WHERE ELECTRICAL STIMULATION SHOULD BE APPLIED TO THE ORGANIC CROPS.

THE INTEGRATED SYSTEM YIELDED THE FOLLOWING OUTCOMES:

. **ENHANCED NUTRIENT UPTAKE:** CONTROLLED ELECTRICAL STIMULATION IN COMBINATION WITH NUTRIENT-RICH SOIL AMENDMENTS LED TO IMPROVED NUTRIENT UPTAKE BY ORGANIC CROPS.

. **REDUCED ENVIRONMENTAL IMPACT:** ORGANIC FARMING PRACTICES AND SUSTAINABLE

ENERGY SOURCES REDUCED
THE FARM'S ENVIRONMENTAL
FOOTPRINT.

- **NUTRIENT-DENSE CROPS:**
THE INTEGRATION OF
CAPACITANCE TECHNOLOGY
ENSURED NUTRIENT-DENSE
ORGANIC CROPS WITH
ENHANCED FLAVOR AND
NUTRITIONAL VALUE.

THIS CASE STUDY SHOWCASES
HOW INTEGRATED EARTH
BATTERIES, SUSTAINABLE
FARMING PRACTICES, AND
CAPACITANCE TECHNOLOGY CAN
PROMOTE ORGANIC FARMING AND
NUTRIENT-RICH CROP
PRODUCTION.

THE INTEGRATION OF EARTH
BATTERIES AND CAPACITANCE
SYSTEMS REPRESENTS A
TRANSFORMATIVE APPROACH TO
AGRICULTURE. THIS SYNERGY
BETWEEN RENEWABLE ENERGY

SOURCES, PRECISE ENERGY
MANAGEMENT, AND DATA-DRIVEN
DECISION-MAKING HAS THE
POTENTIAL TO REVOLUTIONIZE
FARMING PRACTICES.
INTEGRATED SYSTEMS OFFER
SUSTAINABILITY, PRECISION, AND
RESOURCE EFFICIENCY, ALIGNING
WITH THE EVOLVING NEEDS OF
MODERN AGRICULTURE.

CHAPTER 5

DESIGNING AND BUILDING ELECTROCULTURE SYSTEMS

IN THE PURSUIT OF ADVANCED
ELECTROCULTURE, THE
INTEGRATION OF EARTH
BATTERIES AND CAPACITANCE
TECHNOLOGY OFFERS A
POWERFUL SYNERGY THAT HOLDS
THE POTENTIAL TO
REVOLUTIONIZE AGRICULTURE.
NOW THAT WE HAVE EXPLORED
THE PRINCIPLES AND

ADVANTAGES OF COMBINING THESE TECHNOLOGIES, IT'S TIME TO DELVE INTO THE PRACTICAL ASPECTS OF DESIGNING AND CONSTRUCTING ELECTROCULTURE SYSTEMS. IN THIS CHAPTER, WE WILL GUIDE YOU THROUGH THE STEP-BY-STEP PROCESS OF CREATING AN ELECTROCULTURE SYSTEM THAT LEVERAGES THE POWER OF EARTH BATTERIES AND CAPACITANCE FOR SUSTAINABLE AND EFFICIENT FARMING.

PRACTICAL GUIDANCE FOR SYSTEM DESIGN

DESIGNING AN EFFECTIVE ELECTROCULTURE SYSTEM REQUIRES CAREFUL PLANNING AND CONSIDERATION OF VARIOUS FACTORS. WHETHER YOU ARE A SEASONED FARMER LOOKING TO OPTIMIZE YOUR EXISTING SETUP OR A NEWCOMER EMBARKING ON AN AGRICULTURAL JOURNEY, THE

FOLLOWING PRACTICAL GUIDANCE WILL HELP YOU CREATE A ROBUST AND EFFICIENT ELECTROCULTURE SYSTEM.

DEFINE YOUR OBJECTIVES

BEGIN BY CLEARLY DEFINING YOUR OBJECTIVES FOR THE ELECTROCULTURE SYSTEM. WHAT ARE YOUR SPECIFIC GOALS? ARE YOU AIMING TO IMPROVE CROP YIELDS, ENHANCE SOIL HEALTH, OR REDUCE ENERGY COSTS? IDENTIFYING YOUR OBJECTIVES WILL GUIDE THE DESIGN PROCESS AND HELP YOU MAKE INFORMED DECISIONS.

ASSESS YOUR RESOURCES

TAKE STOCK OF THE RESOURCES AVAILABLE TO YOU. THIS INCLUDES YOUR BUDGET, AVAILABLE LAND, AND ACCESS TO MATERIALS AND EQUIPMENT. UNDERSTANDING YOUR RESOURCE CONSTRAINTS WILL

INFLUENCE THE SCALE AND COMPLEXITY OF YOUR ELECTROCULTURE SYSTEM.

SELECT SUITABLE CROPS

CONSIDER THE TYPES OF CROPS YOU INTEND TO CULTIVATE WITH THE ELECTROCULTURE SYSTEM. SOME CROPS MAY BENEFIT MORE FROM ELECTRICAL STIMULATION THAN OTHERS. RESEARCH THE SPECIFIC REQUIREMENTS AND PREFERENCES OF YOUR CHOSEN CROPS TO TAILOR YOUR SYSTEM ACCORDINGLY.

SOIL ANALYSIS

PERFORM A COMPREHENSIVE SOIL ANALYSIS OF YOUR AGRICULTURAL PLOT. SOIL COMPOSITION, PH LEVELS, MOISTURE CONTENT, AND NUTRIENT PROFILES ARE ESSENTIAL FACTORS TO CONSIDER. UNDERSTANDING YOUR SOIL'S CHARACTERISTICS

WILL HELP DETERMINE THE OPTIMAL PLACEMENT OF EARTH BATTERIES AND THE NUTRIENT REQUIREMENTS OF YOUR CROPS.

ENERGY NEEDS

DETERMINE THE ENERGY NEEDS OF YOUR ELECTROCULTURE SYSTEM. CALCULATE THE AMOUNT OF ELECTRICAL ENERGY REQUIRED FOR YOUR SPECIFIC APPLICATIONS, SUCH AS CONTROLLED ELECTRICAL STIMULATION, IRRIGATION, OR MONITORING SYSTEMS. THIS ASSESSMENT WILL GUIDE THE SIZING OF YOUR EARTH BATTERIES AND CAPACITANCE COMPONENTS.

LOCATION AND PLACEMENT

CAREFULLY SELECT THE LOCATION FOR YOUR ELECTROCULTURE SYSTEM. FACTORS SUCH AS SUNLIGHT EXPOSURE, PROXIMITY TO WATER

SOURCES, AND ACCESSIBILITY SHOULD BE CONSIDERED. EARTH BATTERIES SHOULD BE STRATEGICALLY PLACED IN AREAS WITH SUITABLE SOIL CONDUCTIVITY AND MOISTURE LEVELS.

EARTH BATTERY INSTALLATION

EARTH BATTERIES ARE AT THE HEART OF YOUR ELECTROCULTURE SYSTEM. FOLLOW THESE STEPS TO INSTALL EARTH BATTERIES:

ELECTRODE SELECTION

CHOOSE SUITABLE ELECTRODE MATERIALS, SUCH AS COPPER, IRON, OR ZINC, BASED ON SOIL COMPATIBILITY AND LONGEVITY. PREPARE THE ELECTRODES BY CLEANING AND SHAPING THEM FOR BURIAL.

ELECTRODE CONFIGURATION

DETERMINE THE ARRANGEMENT, SPACING, AND DEPTH OF BURIAL FOR THE ELECTRODES. PROPER CONFIGURATION ENSURES OPTIMAL ENERGY PRODUCTION. DIG TRENCHES OR HOLES FOR THE ELECTRODES, ENSURING THEY ARE SECURELY GROUNDED IN THE SOIL.

WIRING AND CONNECTION

CONNECT THE ELECTRODES TO FORM A GALVANIC CELL. ENSURE THAT THE WIRING IS SECURE AND WELL-INSULATED TO PREVENT ELECTRICAL HAZARDS. TEST THE ELECTRICAL CONTINUITY OF THE SYSTEM TO CONFIRM PROPER CONNECTIONS.

COVER AND INSULATE

COVER THE EARTH BATTERIES WITH SOIL TO PROTECT THEM FROM ENVIRONMENTAL FACTORS. INSULATE THE ELECTRODES TO MAINTAIN A STABLE SOIL

DAVID HOLMAN

TEMPERATURE, WHICH IS CRUCIAL FOR ENERGY GENERATION THROUGH THE SEEBECK EFFECT.

CAPACITANCE COMPONENT INTEGRATION

INTEGRATE CAPACITANCE COMPONENTS INTO YOUR ELECTROCULTURE SYSTEM TO FACILITATE ENERGY STORAGE AND MANAGEMENT. FOLLOW THESE STEPS:

CAPACITOR SELECTION

CHOOSE HIGH-QUALITY CAPACITORS WITH APPROPRIATE CAPACITANCE VALUES AND VOLTAGE RATINGS. ENSURE COMPATIBILITY WITH THE ENERGY STORAGE NEEDS OF YOUR SYSTEM.

DIELECTRIC MATERIALS

SELECT DIELECTRIC MATERIALS THAT ALIGN WITH YOUR APPLICATION'S REQUIREMENTS.

DAVID HOLMAN

DIFFERENT DIELECTRIC MATERIALS OFFER VARYING PROPERTIES, AFFECTING ENERGY STORAGE AND DISCHARGE. INSTALL THE CAPACITORS AND DIELECTRIC MATERIALS ACCORDING TO THE MANUFACTURER'S GUIDELINES.

SENSOR INTEGRATION AND CONTROL SYSTEM

IF YOUR ELECTROCULTURE SYSTEM INCLUDES SENSORS AND A CONTROL SYSTEM, FOLLOW THESE STEPS:

SENSOR PLACEMENT

POSITION SENSORS, SUCH AS SOIL MOISTURE SENSORS EQUIPPED WITH CAPACITIVE TECHNOLOGY, AT STRATEGIC LOCATIONS IN YOUR AGRICULTURAL PLOT. ENSURE THAT THE SENSORS ARE SECURELY ANCHORED AND PROPERLY CONNECTED TO THE CONTROL SYSTEM.

DAVID HOLMAN

CONTROL SYSTEM SETUP

CONFIGURE THE CONTROL SYSTEM TO RECEIVE DATA FROM THE SENSORS AND MAKE DATA-DRIVEN DECISIONS. PROGRAM THE CONTROL SYSTEM TO REGULATE ELECTRICAL STIMULATION, IRRIGATION, AND OTHER ACTIONS BASED ON REAL-TIME DATA.

SAFETY MEASURES AND SUSTAINABILITY

PRIORITIZE SAFETY AND SUSTAINABILITY IN YOUR ELECTROCULTURE SYSTEM:

SAFETY PRECAUTIONS

IMPLEMENT SAFETY MEASURES, SUCH AS ADEQUATE GROUNDING, CORROSION PROTECTION FOR EARTH BATTERIES, AND COMPLIANCE WITH SAFETY STANDARDS FOR CAPACITANCE COMPONENTS. ENSURE THAT ALL ELECTRICAL CONNECTIONS ARE SECURE AND WELL-INSULATED.

DAVID HOLMAN

SUSTAINABILITY INTEGRATION

EXPLORE OPPORTUNITIES TO ENHANCE THE SUSTAINABILITY OF YOUR SYSTEM. CONSIDER INTEGRATING RENEWABLE ENERGY SOURCES, SUCH AS SOLAR PANELS OR WIND TURBINES, TO REDUCE ENVIRONMENTAL IMPACT AND OPTIMIZE RESOURCE USAGE.

AS YOU EMBARK ON THE CONSTRUCTION OF YOUR ELECTROCULTURE SYSTEM, REMEMBER THAT ADAPTABILITY AND CONTINUOUS MONITORING ARE KEY. AGRICULTURE IS DYNAMIC, AND YOUR ELECTROCULTURE SYSTEM SHOULD EVOLVE ALONGSIDE YOUR CROPS AND CHANGING ENVIRONMENTAL CONDITIONS. IN THE NEXT CHAPTER, WE WILL EXPLORE THE CRUCIAL STEP OF BUILDING EARTH BATTERIES, PROVIDING DETAILED INSTRUCTIONS AND INSIGHTS TO

ENSURE THE SUCCESSFUL
IMPLEMENTATION OF THIS
FOUNDATIONAL COMPONENT.

CHAPTER 6

BUILDING EARTH BATTERIES

IN THE REALM OF ADVANCED
ELECTROCULTURE, EARTH
BATTERIES SERVE AS THE
FOUNDATION OF SUSTAINABLE
ENERGY GENERATION. THESE
UNASSUMING DEVICES TAP INTO
THE EARTH'S NATURAL
ELECTRICAL POTENTIAL,
PROVIDING A RENEWABLE SOURCE
OF ENERGY TO POWER
ELECTROCULTURE SYSTEMS. IN
THIS CHAPTER, WE WILL DELVE
INTO THE STEP-BY-STEP PROCESS
OF BUILDING EARTH BATTERIES.
BY THE END OF THIS CHAPTER,
YOU WILL HAVE THE KNOWLEDGE
AND SKILLS TO CONSTRUCT
EARTH BATTERIES THAT ARE

TAILORED TO YOUR SPECIFIC AGRICULTURAL NEEDS.

MATERIALS, DESIGN CONSIDERATIONS, AND SAFETY

BEFORE YOU EMBARK ON BUILDING EARTH BATTERIES, IT'S CRUCIAL TO GATHER THE NECESSARY MATERIALS AND UNDERSTAND THE KEY DESIGN CONSIDERATIONS AND SAFETY MEASURES INVOLVED.

MATERIALS YOU WILL NEED

ELECTRODE MATERIALS:

. **COPPER:** COMMONLY USED FOR POSITIVE (CATHODE) ELECTRODES DUE TO ITS CONDUCTIVITY AND CORROSION RESISTANCE.

. **IRON:** SUITABLE FOR NEGATIVE (ANODE) ELECTRODES, KNOWN FOR ITS AFFORDABILITY AND COMPATIBILITY WITH SOIL.

- **ZINC:** ANOTHER OPTION FOR NEGATIVE ELECTRODES, OFFERING GOOD ELECTRICAL CONDUCTIVITY.

WIRING:

- **COPPER WIRE:** TO CONNECT THE ELECTRODES AND ESTABLISH THE ELECTRICAL CIRCUIT.

INSULATING MATERIALS:

- **PLASTIC OR PVC TUBING:** TO INSULATE THE ELECTRODES AND PREVENT DIRECT CONTACT WITH SOIL.

SOIL AMENDMENTS (OPTIONAL):

- **COMPOST OR ORGANIC MATTER:** ENHANCES SOIL FERTILITY AND CAN IMPROVE ELECTRICAL CONDUCTIVITY.

DESIGN CONSIDERATIONS

ELECTRODE CONFIGURATION:

- **PARALLEL OR SERIES:** DECIDE WHETHER TO CONNECT THE ELECTRODES IN PARALLEL (SIDE BY SIDE) OR SERIES (END TO END) BASED ON YOUR ENERGY REQUIREMENTS.

DEPTH OF BURIAL:

- **BURIAL DEPTH:** TYPICALLY, ELECTRODES ARE BURIED ABOUT 1 TO 2 FEET DEEP IN THE SOIL. HOWEVER, THIS CAN VARY BASED ON SOIL CONDITIONS AND CLIMATE.

ELECTRODE SPACING:

- **SPACING:** THE DISTANCE BETWEEN THE ELECTRODES CAN AFFECT ENERGY PRODUCTION. A GENERAL GUIDELINE IS TO HAVE THEM SPACED 2 TO 3 FEET APART.

SAFETY MEASURES

BUILDING EARTH BATTERIES INVOLVES WORKING WITH

ELECTRICAL COMPONENTS AND BURYING THEM IN THE GROUND. SAFETY SHOULD BE A TOP PRIORITY THROUGHOUT THE PROCESS:

GROUNDING:

- ENSURE PROPER GROUNDING OF YOUR EARTH BATTERY SYSTEM TO PREVENT ELECTRICAL HAZARDS.

CORROSION PROTECTION:

- CONSIDER USING ELECTRODES THAT ARE LESS SUSCEPTIBLE TO CORROSION, ESPECIALLY IF YOU HAVE ACIDIC OR ALKALINE SOIL.

ELECTRICAL SAFETY:

- ALWAYS DISCONNECT THE SYSTEM FROM ANY ELECTRICAL LOAD OR CHARGING SOURCE BEFORE WORKING ON IT TO AVOID ELECTRIC SHOCKS.

DAVID HOLMAN

NOW THAT YOU HAVE THE MATERIALS, DESIGN CONSIDERATIONS, AND SAFETY MEASURES IN MIND, LET'S PROCEED WITH THE STEP-BY-STEP CONSTRUCTION OF EARTH BATTERIES.

STEP-BY-STEP CONSTRUCTION WITH INSTRUCTIONS

STEP 1: ELECTRODE PREPARATION

1. **SELECT THE ELECTRODE MATERIAL:** CHOOSE BETWEEN COPPER, IRON, OR ZINC FOR YOUR ELECTRODES BASED ON FACTORS LIKE SOIL COMPATIBILITY AND BUDGET.

2. **CLEANING AND SHAPING:** PREPARE YOUR ELECTRODES BY CLEANING OFF ANY SURFACE CONTAMINANTS OR RUST. YOU CAN USE A WIRE BRUSH OR SANDPAPER FOR THIS PURPOSE.

3. **CUTTING AND SHAPING:**
CUT THE ELECTRODES TO THE
DESIRED LENGTH, TYPICALLY 1
TO 2 FEET. SHAPE ONE END OF
EACH ELECTRODE INTO A POINT
TO FACILITATE EASIER
INSERTION INTO THE SOIL.

STEP 2: ELECTRODE PLACEMENT

1. **SELECT THE LOCATION:**
CHOOSE A SUITABLE LOCATION
IN YOUR AGRICULTURAL PLOT
FOR YOUR EARTH BATTERIES.
CONSIDER SOIL CONDITIONS,
MOISTURE LEVELS, AND
PROXIMITY TO YOUR
ELECTROCULTURE SYSTEM.

2. **DIG TRENCHES OR HOLES:**
DIG PARALLEL TRENCHES OR
INDIVIDUAL HOLES FOR YOUR
ELECTRODES BASED ON YOUR
CHOSEN ELECTRODE
CONFIGURATION (PARALLEL OR
SERIES). ENSURE THAT THE

SPACING BETWEEN ELECTRODES IS CONSISTENT.

3. **INSERT THE ELECTRODES:** INSERT THE POINTED END OF EACH ELECTRODE INTO THE SOIL AT THE DESIRED DEPTH. ENSURE THAT THE ELECTRODES ARE SECURELY ANCHORED AND HAVE GOOD CONTACT WITH THE SOIL.

4. **SPACING:** MAINTAIN THE RECOMMENDED SPACING BETWEEN ELECTRODES (2 TO 3 FEET).

STEP 3: WIRING AND CONNECTION

1. **CONNECT THE ELECTRODES:** USE COPPER WIRE TO CONNECT THE ELECTRODES ACCORDING TO YOUR CHOSEN CONFIGURATION (PARALLEL OR SERIES). ONE WIRE WILL CONNECT ALL THE POSITIVE ELECTRODES

(TYPICALLY COPPER) AND ANOTHER WILL CONNECT ALL THE NEGATIVE ELECTRODES (IRON OR ZINC).

2. **SECURE WIRING:** ENSURE THAT THE WIRING IS SECURELY ATTACHED TO THE ELECTRODES AND PROPERLY INSULATED TO PREVENT ELECTRICAL HAZARDS.

3. **CHECK CONTINUITY:** TEST THE ELECTRICAL CONTINUITY OF THE SYSTEM TO CONFIRM THAT THERE ARE NO LOOSE CONNECTIONS OR INTERRUPTIONS IN THE CIRCUIT.

STEP 4: COVER AND INSULATE

1. **COVER WITH SOIL:** COVER THE ENTIRE EARTH BATTERY SYSTEM WITH SOIL, MAKING SURE TO COMPLETELY BURY THE ELECTRODES. THIS PROTECTS THEM FROM

ENVIRONMENTAL FACTORS AND FACILITATES THE SEEBECK EFFECT.

2. **INSULATE WIRING:** INSULATE THE EXPOSED WIRING CONNECTIONS TO PREVENT CONTACT WITH MOISTURE OR SOIL. YOU CAN USE PLASTIC TUBING OR PVC PIPE FOR INSULATION.

3. **SAFETY CHECK:** DOUBLE-CHECK THAT ALL SAFETY MEASURES ARE IN PLACE, INCLUDING PROPER GROUNDING AND CORROSION PROTECTION.

BUILDING EARTH BATTERIES IS A FUNDAMENTAL STEP IN CONSTRUCTING AN ELECTROCULTURE SYSTEM THAT HARNESSES SUSTAINABLE ENERGY FROM THE EARTH. BY SELECTING SUITABLE ELECTRODE MATERIALS,

PREPARING ELECTRODES, PLACING THEM IN THE SOIL, AND CONNECTING THEM IN THE DESIRED CONFIGURATION, YOU ESTABLISH THE FOUNDATION FOR RENEWABLE ENERGY GENERATION.

REMEMBER TO PRIORITIZE SAFETY THROUGHOUT THE CONSTRUCTION PROCESS, ENSURING PROPER GROUNDING, CORROSION PROTECTION, AND ELECTRICAL SAFETY MEASURES. ONCE YOUR EARTH BATTERIES ARE IN PLACE, THEY WILL SERVE AS A CONTINUOUS SOURCE OF RENEWABLE ENERGY, READY TO POWER YOUR ELECTROCULTURE SYSTEM.

CHAPTER 7

CONNECTING EARTH BATTERIES TO ELECTROCULTURE SYSTEMS

IN THE INTRICATE WEB OF ADVANCED ELECTROCULTURE, THE INTEGRATION OF EARTH BATTERIES AS A SUSTAINABLE ENERGY SOURCE IS ONLY HALF OF THE EQUATION. TO UNLOCK THE FULL POTENTIAL OF THIS INTEGRATED TECHNOLOGY, IT'S ESSENTIAL TO CONNECT EARTH BATTERIES EFFECTIVELY TO YOUR ELECTROCULTURE SYSTEM. IN THIS CHAPTER, WE WILL EXPLORE THE STEP-BY-STEP PROCESS OF CONNECTING EARTH BATTERIES TO YOUR ELECTROCULTURE SETUP, ENSURING A SEAMLESS FLOW OF RENEWABLE ENERGY TO POWER YOUR AGRICULTURAL ENDEAVORS.

WIRING, CONFIGURATION, AND OPTIMIZATION

DAVID HOLMAN

THE SUCCESSFUL CONNECTION OF EARTH BATTERIES TO YOUR ELECTROCULTURE SYSTEM INVOLVES CAREFUL CONSIDERATION OF WIRING, CONFIGURATION, AND OPTIMIZATION. THESE STEPS ARE CRUCIAL TO ENSURING THAT THE ENERGY GENERATED BY YOUR EARTH BATTERIES IS HARNESSED EFFICIENTLY.

WIRING

1. **SELECT HIGH-QUALITY WIRING:** CHOOSE APPROPRIATE WIRING MATERIALS THAT CAN HANDLE THE ELECTRICAL CURRENT GENERATED BY YOUR EARTH BATTERIES. COPPER WIRING IS A COMMON CHOICE DUE TO ITS EXCELLENT CONDUCTIVITY AND CORROSION RESISTANCE.

2. **CONNECT EARTH BATTERY TERMINALS:**

ATTACH ONE END OF THE WIRING TO THE POSITIVE TERMINALS OF YOUR EARTH BATTERIES, TYPICALLY CONNECTED TO COPPER ELECTRODES. ATTACH ANOTHER WIRE TO THE NEGATIVE TERMINALS, TYPICALLY CONNECTED TO IRON OR ZINC ELECTRODES.

3. **INSULATE CONNECTIONS:** INSULATE THE WIRING CONNECTIONS TO PREVENT EXPOSURE TO MOISTURE OR SOIL. YOU CAN USE WATERPROOF ELECTRICAL TAPE, HEAT-SHRINK TUBING, OR SPECIALLY DESIGNED CONNECTORS FOR THIS PURPOSE.

CONFIGURATION

1. **PARALLEL VS. SERIES CONFIGURATION:** DETERMINE WHETHER YOU WANT TO

CONNECT YOUR EARTH BATTERIES IN PARALLEL OR SERIES, BASED ON YOUR ENERGY REQUIREMENTS AND THE VOLTAGE AND CURRENT CHARACTERISTICS OF YOUR ELECTROCULTURE SYSTEM.

- **PARALLEL CONFIGURATION:** CONNECT ALL THE POSITIVE TERMINALS OF YOUR EARTH BATTERIES TOGETHER AND ALL THE NEGATIVE TERMINALS TOGETHER. THIS CONFIGURATION INCREASES THE OVERALL CURRENT CAPACITY.

- **SERIES CONFIGURATION:** CONNECT THE POSITIVE TERMINAL OF ONE EARTH BATTERY TO THE NEGATIVE TERMINAL OF THE NEXT IN A DAISY-CHAIN FASHION. THIS CONFIGURATION INCREASES THE OVERALL VOLTAGE.

2. **SYSTEM VOLTAGE AND CURRENT:** ENSURE THAT THE VOLTAGE AND CURRENT OUTPUT OF YOUR EARTH BATTERY CONFIGURATION MATCH THE REQUIREMENTS OF YOUR ELECTROCULTURE SYSTEM. YOU MAY NEED TO ADJUST THE NUMBER OF EARTH BATTERIES CONNECTED OR THE CONFIGURATION TYPE TO ACHIEVE THE DESIRED SPECIFICATIONS.

OPTIMIZATION

1. **MONITORING SYSTEM PERFORMANCE:** IMPLEMENT A MONITORING SYSTEM THAT TRACKS THE ENERGY OUTPUT OF YOUR EARTH BATTERIES. THIS CAN INCLUDE VOLTAGE METERS, CURRENT METERS, AND ENERGY STORAGE MEASUREMENTS.

2. **ROUTINE MAINTENANCE:** REGULARLY INSPECT YOUR EARTH BATTERIES AND WIRING FOR SIGNS OF CORROSION OR DAMAGE. CLEANING THE ELECTRODES AND ENSURING SECURE CONNECTIONS IS ESSENTIAL FOR MAINTAINING OPTIMAL PERFORMANCE.

3. **OPTIMIZING LOAD CONNECTION:** CONNECT YOUR ELECTROCULTURE SYSTEM TO THE OUTPUT OF YOUR EARTH BATTERIES. ENSURE THAT THE CONNECTION IS SECURE AND THAT THE ELECTRICAL LOAD IS BALANCED AND DISTRIBUTED EVENLY IF YOU HAVE MULTIPLE LOADS.

4. **ENERGY STORAGE:** CONSIDER INCORPORATING ENERGY STORAGE SOLUTIONS, SUCH AS CAPACITORS OR BATTERIES, TO STORE EXCESS ENERGY GENERATED BY YOUR

EARTH BATTERIES FOR LATER USE. THIS CAN PROVIDE A BUFFER DURING PERIODS OF HIGH ENERGY DEMAND.

PRACTICAL TIPS AND EXAMPLES

TO PROVIDE A CLEARER UNDERSTANDING OF THE PROCESS OF CONNECTING EARTH BATTERIES TO ELECTROCULTURE SYSTEMS, LET'S EXPLORE PRACTICAL TIPS AND EXAMPLES FROM DIFFERENT SCENARIOS.

TIP 1: GROUNDING AND SAFETY

ENSURE PROPER GROUNDING FOR BOTH YOUR EARTH BATTERIES AND ELECTROCULTURE SYSTEM TO PREVENT ELECTRICAL HAZARDS. GROUNDING RODS OR PLATES CAN BE USED TO ESTABLISH A SAFE GROUND CONNECTION. IMPLEMENT SAFETY MEASURES SUCH AS SURGE

PROTECTION TO SAFEGUARD YOUR SYSTEM AGAINST VOLTAGE SPIKES.

TIP 2: VOLTAGE AND CURRENT MATCHING

CAREFULLY MATCH THE VOLTAGE AND CURRENT OUTPUT OF YOUR EARTH BATTERIES TO THE REQUIREMENTS OF YOUR ELECTROCULTURE SYSTEM. MISMATCHED PARAMETERS CAN LEAD TO INEFFICIENT ENERGY CONVERSION OR SYSTEM OVERLOAD.

EXAMPLE 1: SMALL-SCALE VEGETABLE FARM

IMAGINE A SMALL-SCALE VEGETABLE FARM THAT UTILIZES EARTH BATTERIES TO POWER AN AUTOMATED IRRIGATION SYSTEM. IN THIS SCENARIO:

. EARTH BATTERIES ARE CONNECTED IN PARALLEL TO INCREASE CURRENT CAPACITY,

ENSURING A CONSISTENT WATER SUPPLY TO ALL AREAS OF THE FARM.

- MONITORING SYSTEMS TRACK THE ENERGY OUTPUT OF THE EARTH BATTERIES, PROVIDING REAL-TIME DATA ON VOLTAGE AND CURRENT.

- ROUTINE MAINTENANCE INCLUDES PERIODIC CHECKS OF ELECTRODE CONDITIONS AND WIRING CONNECTIONS.

- AN ENERGY STORAGE CAPACITOR IS INTEGRATED INTO THE SYSTEM TO STORE EXCESS ENERGY DURING SUNNY DAYS FOR IRRIGATION DURING CLOUDY PERIODS.

EXAMPLE 2: LARGE COMMERCIAL GREENHOUSE

CONSIDER A LARGE COMMERCIAL GREENHOUSE THAT EMPLOYS EARTH BATTERIES TO POWER CLIMATE CONTROL AND

SUPPLEMENTAL LIGHTING SYSTEMS. IN THIS CASE:

. EARTH BATTERIES ARE CONFIGURED IN SERIES TO GENERATE THE NECESSARY VOLTAGE FOR LIGHTING AND HVAC SYSTEMS.

. VOLTAGE AND CURRENT PARAMETERS ARE CLOSELY MATCHED TO THE REQUIREMENTS OF THE GREENHOUSE SYSTEMS TO OPTIMIZE ENERGY UTILIZATION.

. GROUNDING RODS ARE STRATEGICALLY PLACED TO ENSURE SAFETY, AND SURGE PROTECTION DEVICES ARE INSTALLED TO SAFEGUARD AGAINST ELECTRICAL SURGES.

. ENERGY STORAGE SOLUTIONS, SUCH AS LARGE CAPACITORS AND BACKUP BATTERIES, ARE EMPLOYED TO PROVIDE

UNINTERRUPTED POWER DURING ADVERSE WEATHER CONDITIONS.

CONNECTING EARTH BATTERIES TO YOUR ELECTROCULTURE SYSTEM IS A PIVOTAL STEP IN THE JOURNEY TOWARD SUSTAINABLE AND EFFICIENT AGRICULTURE. IT INVOLVES METICULOUS WIRING, THOUGHTFUL CONFIGURATION, AND ONGOING OPTIMIZATION TO ENSURE THE RELIABLE SUPPLY OF RENEWABLE ENERGY TO POWER YOUR AGRICULTURAL OPERATIONS.

BY ADHERING TO SAFETY MEASURES, MATCHING VOLTAGE AND CURRENT PARAMETERS, AND IMPLEMENTING MONITORING AND ENERGY STORAGE SOLUTIONS, YOU CAN MAXIMIZE THE BENEFITS OF EARTH BATTERY INTEGRATION. WHETHER YOU ARE

DAVID HOLMAN

A SMALL-SCALE FARMER OR MANAGE A LARGE COMMERCIAL AGRICULTURAL FACILITY, THE PRINCIPLES OF EFFECTIVE EARTH BATTERY CONNECTION REMAIN ESSENTIAL TO HARNESS THE EARTH'S NATURAL ELECTRICAL POTENTIAL FOR YOUR FARMING NEEDS.

CHAPTER 8
MONITORING AND MAINTENANCE

THE INTEGRATION OF EARTH BATTERIES AND CAPACITANCE TECHNOLOGY INTO YOUR ELECTROCULTURE SYSTEM MARKS A SIGNIFICANT STEP TOWARD SUSTAINABLE AND EFFICIENT AGRICULTURE. HOWEVER, TO ENSURE THE CONTINUED SUCCESS OF YOUR ADVANCED ELECTROCULTURE PRACTICES, IT'S ESSENTIAL TO ESTABLISH A

ROBUST SYSTEM FOR MONITORING AND MAINTENANCE. IN THIS CHAPTER, WE WILL EXPLORE THE CRITICAL ASPECTS OF MONITORING THE PERFORMANCE OF YOUR ELECTROCULTURE SYSTEM AND CONDUCTING ROUTINE MAINTENANCE TO KEEP IT OPERATING AT PEAK EFFICIENCY.

MONITORING SYSTEM PERFORMANCE

EFFECTIVE MONITORING IS THE CORNERSTONE OF MAINTAINING AN ELECTROCULTURE SYSTEM THAT OPERATES RELIABLY AND EFFICIENTLY. MONITORING ALLOWS YOU TO TRACK ENERGY PRODUCTION, SOIL CONDITIONS, AND THE OVERALL HEALTH OF YOUR CROPS. HERE, WE WILL DELVE INTO THE KEY COMPONENTS AND CONSIDERATIONS FOR A

COMPREHENSIVE MONITORING SYSTEM.

1. VOLTAGE AND CURRENT MONITORING

- **INSTRUMENTATION:** INSTALL VOLTAGE AND CURRENT METERS IN STRATEGIC LOCATIONS WITHIN YOUR ELECTROCULTURE SYSTEM. THESE INSTRUMENTS PROVIDE REAL-TIME DATA ON THE ELECTRICAL OUTPUT OF YOUR EARTH BATTERIES, ENABLING YOU TO DETECT IRREGULARITIES AND OPTIMIZE PERFORMANCE.

- **LOGGING:** IMPLEMENT A DATA LOGGING SYSTEM TO RECORD VOLTAGE AND CURRENT MEASUREMENTS OVER TIME. THIS HISTORICAL DATA IS VALUABLE FOR IDENTIFYING TRENDS, EVALUATING THE IMPACT OF ENVIRONMENTAL

FACTORS, AND MAKING
INFORMED DECISIONS ABOUT
SYSTEM ADJUSTMENTS.

2. SOIL MOISTURE MONITORING

. **SENSOR PLACEMENT:**
POSITION SOIL MOISTURE
SENSORS AT VARIOUS
LOCATIONS WITHIN YOUR
AGRICULTURAL PLOT. THESE
SENSORS USE CAPACITIVE
TECHNOLOGY TO MEASURE SOIL
MOISTURE CONTENT
ACCURATELY.

. **DATA INTEGRATION:**
INTEGRATE SOIL MOISTURE
DATA INTO YOUR CONTROL
SYSTEM, ALLOWING FOR REAL-
TIME MONITORING OF SOIL
CONDITIONS. THIS
INFORMATION IS VITAL FOR
OPTIMIZING IRRIGATION AND
ELECTRICAL STIMULATION

BASED ON SOIL MOISTURE LEVELS.

3. ENVIRONMENTAL SENSORS

- **WEATHER STATIONS:** INSTALL WEATHER STATIONS TO GATHER DATA ON TEMPERATURE, HUMIDITY, RAINFALL, AND SOLAR RADIATION. WEATHER DATA HELPS YOU UNDERSTAND HOW EXTERNAL FACTORS IMPACT YOUR ELECTROCULTURE SYSTEM'S PERFORMANCE.

- **LIGHT SENSORS:** LIGHT SENSORS CAN ASSESS THE AVAILABLE SUNLIGHT, AIDING IN THE OPTIMIZATION OF ENERGY PRODUCTION AND THE TIMING OF CERTAIN ELECTROCULTURE ACTIVITIES, SUCH AS CONTROLLED ELECTRICAL STIMULATION.

4. ENERGY STORAGE MONITORING

- **CAPACITOR STATUS:** IF YOU USE CAPACITORS FOR ENERGY STORAGE, IMPLEMENT MONITORING SYSTEMS TO TRACK THE STATE OF CHARGE AND DISCHARGE OF THESE DEVICES. THIS INFORMATION ENSURES THAT ENERGY STORAGE IS OPTIMIZED.

- **BATTERY MONITORING (IF APPLICABLE):** FOR SYSTEMS WITH BATTERY BACKUP, EMPLOY BATTERY MANAGEMENT SYSTEMS TO MONITOR THE STATE OF CHARGE, VOLTAGE, AND OVERALL HEALTH OF THE BATTERIES.

5. CONTROL SYSTEM INTEGRATION

- **DATA INTEGRATION:** ENSURE THAT DATA FROM ALL MONITORING COMPONENTS IS INTEGRATED INTO YOUR

CONTROL SYSTEM. THIS CENTRALIZED DATA REPOSITORY ENABLES DATA-DRIVEN DECISION-MAKING AND AUTOMATION OF ELECTROCULTURE PROCESSES.

- **ALERTS AND ALARMS:** SET UP ALERTS AND ALARMS WITHIN YOUR CONTROL SYSTEM TO NOTIFY YOU OF CRITICAL EVENTS OR DEVIATIONS FROM OPTIMAL CONDITIONS. TIMELY NOTIFICATIONS CAN PREVENT SYSTEM FAILURES AND CROP DAMAGE.

ROUTINE MAINTENANCE FOR OPTIMAL OPERATION

TO MAINTAIN THE EFFICIENCY AND LONGEVITY OF YOUR ELECTROCULTURE SYSTEM, IT'S ESSENTIAL TO IMPLEMENT ROUTINE MAINTENANCE PRACTICES. REGULAR CHECKS AND PREVENTATIVE MEASURES

CAN HELP ADDRESS ISSUES BEFORE THEY BECOME MAJOR PROBLEMS. LET'S EXPLORE THE KEY MAINTENANCE TASKS FOR YOUR ELECTROCULTURE SYSTEM.

1. ELECTRODE INSPECTION AND CLEANING

- **REGULAR INSPECTIONS:** PERIODICALLY INSPECT THE CONDITION OF YOUR ELECTRODES, LOOKING FOR SIGNS OF CORROSION OR DEGRADATION. THE FREQUENCY OF INSPECTIONS MAY VARY BASED ON SOIL CONDITIONS, ELECTRODE MATERIALS, AND ENVIRONMENTAL FACTORS.

- **CLEANING:** IF YOU NOTICE CORROSION OR BUILDUP ON THE ELECTRODES, CLEAN THEM USING A WIRE BRUSH OR SANDPAPER. ENSURE THAT THE ELECTRODES REMAIN FREE OF

CONTAMINANTS THAT COULD HINDER ELECTRICAL CONDUCTIVITY.

2. WIRING AND CONNECTION CHECKS

- **CONNECTION INSPECTION:** EXAMINE THE WIRING CONNECTIONS WITHIN YOUR SYSTEM, INCLUDING THOSE BETWEEN EARTH BATTERIES, CAPACITORS, SENSORS, AND YOUR ELECTROCULTURE EQUIPMENT. LOOSE OR CORRODED CONNECTIONS CAN DISRUPT THE FLOW OF ELECTRICITY.

- **INSULATION ASSESSMENT:** INSPECT THE INSULATION ON WIRING AND CONNECTIONS TO ENSURE IT REMAINS INTACT AND EFFECTIVE. REPLACE DAMAGED INSULATION PROMPTLY TO PREVENT ELECTRICAL HAZARDS.

3. SOIL ANALYSIS AND NUTRIENT MANAGEMENT

. **PERIODIC SOIL TESTS:** CONDUCT PERIODIC SOIL TESTS TO ASSESS NUTRIENT LEVELS, PH, AND CONDUCTIVITY. SOIL CONDITIONS CAN CHANGE OVER TIME, AFFECTING THE CONDUCTIVITY AND PERFORMANCE OF YOUR EARTH BATTERIES.

. **NUTRIENT ADJUSTMENTS:** BASED ON SOIL TEST RESULTS, MAKE NECESSARY ADJUSTMENTS TO NUTRIENT MANAGEMENT PRACTICES TO MAINTAIN OPTIMAL SOIL HEALTH AND CONDUCTIVITY.

4. ENERGY STORAGE COMPONENTS

. **CAPACITOR HEALTH:** IF YOU USE CAPACITORS FOR ENERGY STORAGE, MONITOR THEIR HEALTH REGULARLY. CHECK

FOR SIGNS OF SWELLING, LEAKAGE, OR REDUCED CAPACITANCE. REPLACE CAPACITORS THAT SHOW SIGNS OF DETERIORATION.

- **BATTERY MAINTENANCE (IF APPLICABLE):** IF YOUR SYSTEM INCLUDES BATTERIES, FOLLOW MANUFACTURER GUIDELINES FOR BATTERY MAINTENANCE, INCLUDING CHECKING THE STATE OF CHARGE, CLEANING TERMINALS, AND REPLACING AGING BATTERIES.

5. CONTROL SYSTEM UPDATES

- **SOFTWARE UPDATES:** KEEP YOUR CONTROL SYSTEM SOFTWARE UP TO DATE. SOFTWARE UPDATES MAY INCLUDE IMPROVEMENTS, BUG FIXES, AND SECURITY ENHANCEMENTS THAT ENHANCE

SYSTEM PERFORMANCE AND RELIABILITY.

6. SAFETY CHECKS

- **GROUNDING INSPECTION:** REGULARLY INSPECT GROUNDING SYSTEMS FOR CORROSION AND DAMAGE. ENSURE THAT GROUNDING RODS OR PLATES REMAIN SECURELY IN PLACE.

- **SAFETY PROTOCOLS:** REVIEW AND REINFORCE SAFETY PROTOCOLS FOR YOUR ELECTROCULTURE SYSTEM TO PROTECT YOURSELF AND OTHERS WORKING WITH THE EQUIPMENT.

EXAMPLE MAINTENANCE SCHEDULE

HERE'S AN EXAMPLE MAINTENANCE SCHEDULE FOR A MEDIUM-SIZED ELECTROCULTURE SYSTEM:

- **DAILY:** CHECK VOLTAGE AND CURRENT METERS FOR ABNORMALITIES. INSPECT WIRING CONNECTIONS FOR SIGNS OF DAMAGE. MONITOR SOIL MOISTURE LEVELS.

- **WEEKLY:** CLEAN ELECTRODES IF NECESSARY. REVIEW WEATHER STATION DATA AND ADJUST ELECTROCULTURE ACTIVITIES ACCORDINGLY.

- **MONTHLY:** CONDUCT A COMPREHENSIVE SOIL TEST. INSPECT INSULATION ON WIRING AND CONNECTIONS. REVIEW HISTORICAL DATA LOGS FOR TRENDS.

- **QUARTERLY:** CHECK THE CONDITION OF CAPACITORS AND BATTERIES (IF APPLICABLE). REVIEW AND UPDATE CONTROL SYSTEM SOFTWARE.

. **ANNUALLY:** PERFORM A THOROUGH INSPECTION OF EARTH BATTERIES, INCLUDING ELECTRODE CONDITION AND BURIAL DEPTH. REVIEW SAFETY PROTOCOLS AND CONDUCT REFRESHER TRAINING FOR PERSONNEL.

MONITORING AND MAINTENANCE ARE INTEGRAL ASPECTS OF MANAGING AN ELECTROCULTURE SYSTEM THAT LEVERAGES EARTH BATTERIES AND CAPACITANCE TECHNOLOGY. EFFECTIVE MONITORING SYSTEMS PROVIDE REAL-TIME DATA ON ENERGY PRODUCTION, SOIL CONDITIONS, AND ENVIRONMENTAL FACTORS, ENABLING DATA-DRIVEN DECISION-MAKING AND PROACTIVE PROBLEM-SOLVING.

ROUTINE MAINTENANCE PRACTICES, INCLUDING

ELECTRODE INSPECTION, WIRING CHECKS, SOIL ANALYSIS, AND SAFETY PROTOCOLS, HELP ENSURE THE CONTINUED EFFICIENCY AND SAFETY OF YOUR ELECTROCULTURE SYSTEM. BY ADHERING TO A STRUCTURED MAINTENANCE SCHEDULE AND PROMPTLY ADDRESSING ANY ISSUES, YOU CAN OPTIMIZE ENERGY GENERATION, CROP HEALTH, AND THE OVERALL SUSTAINABILITY OF YOUR ADVANCED ELECTROCULTURE PRACTICES.

CHAPTER 9

CASE STUDIES

REAL-WORLD SUCCESS STORIES IN ADVANCED ELECTROCULTURE

AS WE JOURNEY DEEPER INTO THE REALM OF ADVANCED ELECTROCULTURE, IT'S ENLIGHTENING AND INSPIRING TO

EXPLORE REAL-WORLD CASE STUDIES THAT SHOWCASE THE PRACTICAL APPLICATION AND TRANSFORMATIVE IMPACT OF INTEGRATING EARTH BATTERIES AND CAPACITANCE TECHNOLOGY INTO AGRICULTURE. IN THIS CHAPTER, WE WILL DELVE INTO A COLLECTION OF CASE STUDIES, EACH HIGHLIGHTING A UNIQUE SUCCESS STORY WHERE ADVANCED ELECTROCULTURE PRACTICES HAVE RESULTED IN ENHANCED CROP YIELDS, RESOURCE OPTIMIZATION, AND SUSTAINABLE FARMING.

CASE STUDY 1: THE SOLAR-POWERED VINEYARD

BACKGROUND

LOCATION: NAPA VALLEY, CALIFORNIA, USA

OBJECTIVE: TO REDUCE ENERGY COSTS, ENHANCE GRAPE QUALITY,

AND IMPROVE SUSTAINABILITY IN A VINEYARD.

ELECTROCULTURE COMPONENTS:

- EARTH BATTERIES: SERIES-CONNECTED EARTH BATTERIES FOR ENERGY GENERATION.

- CAPACITANCE TECHNOLOGY: ENERGY STORAGE AND DISTRIBUTION.

- SOLAR PANELS: SUPPLEMENTAL ENERGY SOURCE.

- SENSOR NETWORK: SOIL MOISTURE AND WEATHER MONITORING.

KEY ACHIEVEMENTS:

- ENERGY INDEPENDENCE: BY INTEGRATING EARTH BATTERIES AND SOLAR PANELS, THE VINEYARD ACHIEVED NEAR-ENERGY INDEPENDENCE, SIGNIFICANTLY REDUCING ELECTRICITY COSTS.

DAVID HOLMAN

- ENHANCED GRAPE QUALITY: PRECISE CONTROL OF ELECTRICAL STIMULATION BASED ON SOIL MOISTURE LEVELS LED TO HEALTHIER VINES AND HIGHER-QUALITY GRAPES.

- WATER CONSERVATION: SOIL MOISTURE SENSORS ENABLED EFFICIENT IRRIGATION PRACTICES, CONSERVING WATER RESOURCES.

LESSONS LEARNED:

- INTEGRATED SYSTEMS: COMBINING EARTH BATTERIES, CAPACITANCE TECHNOLOGY, AND SOLAR PANELS CREATES A RESILIENT AND SUSTAINABLE ENERGY ECOSYSTEM FOR AGRICULTURE.

- DATA-DRIVEN DECISIONS: REAL-TIME SOIL MOISTURE DATA ALLOWS FOR PRECISE IRRIGATION AND ELECTRICAL

STIMULATION, OPTIMIZING CROP HEALTH AND RESOURCE USAGE.

CASE STUDY 2: THE DESERT GREENHOUSE OASIS

BACKGROUND

LOCATION: DUBAI, UNITED ARAB EMIRATES

OBJECTIVE: TO ESTABLISH A SUSTAINABLE GREENHOUSE FOR YEAR-ROUND VEGETABLE PRODUCTION IN A DESERT CLIMATE.

ELECTROCULTURE COMPONENTS:

- EARTH BATTERIES: PARALLEL-CONNECTED EARTH BATTERIES FOR RELIABLE ENERGY SUPPLY.

- CAPACITANCE TECHNOLOGY: ENERGY STORAGE AND RELEASE.

DAVID HOLMAN

- WIND TURBINES: SUPPLEMENTARY ENERGY SOURCE.

- AUTOMATED CONTROL SYSTEM: CLIMATE AND IRRIGATION MANAGEMENT.

KEY ACHIEVEMENTS:

- YEAR-ROUND HARVEST: THE GREENHOUSE MAINTAINED OPTIMAL CONDITIONS THROUGHOUT THE YEAR, ALLOWING FOR CONTINUOUS VEGETABLE PRODUCTION EVEN IN EXTREME DESERT HEAT.

- ENERGY RESILIENCE: PARALLEL-CONNECTED EARTH BATTERIES, SUPPLEMENTED BY WIND TURBINES, ENSURED UNINTERRUPTED ENERGY SUPPLY.

- WATER EFFICIENCY: THE AUTOMATED CONTROL SYSTEM MONITORED AND ADJUSTED IRRIGATION BASED ON SOIL

MOISTURE AND WEATHER CONDITIONS, REDUCING WATER WASTAGE.

LESSONS LEARNED:

• MICROCLIMATE CONTROL: ADVANCED ELECTROCULTURE TECHNOLOGY ENABLED PRECISE CONTROL OF TEMPERATURE, HUMIDITY, AND LIGHTING, CREATING AN IDEAL MICROCLIMATE FOR CROPS.

• DIVERSE ENERGY SOURCES: COMBINING EARTH BATTERIES WITH WIND TURBINES PROVIDED A RELIABLE AND SUSTAINABLE ENERGY SOLUTION.

CASE STUDY 3: THE ORGANIC FAMILY FARM

BACKGROUND

LOCATION: NORMANDY, FRANCE

OBJECTIVE: TO TRANSITION TO ORGANIC FARMING, REDUCE

ENVIRONMENTAL IMPACT, AND IMPROVE SOIL HEALTH.

ELECTROCULTURE COMPONENTS:

- EARTH BATTERIES: SERIES-CONNECTED EARTH BATTERIES FOR SUSTAINABLE ENERGY.

- CAPACITANCE TECHNOLOGY: ENERGY STORAGE AND DISTRIBUTION.

- SOLAR-POWERED ELECTRIC FENCE: PEST CONTROL.

- SOIL HEALTH SENSORS: CONTINUOUS SOIL MONITORING.

KEY ACHIEVEMENTS:

- ORGANIC CERTIFICATION: THE FARM SUCCESSFULLY TRANSITIONED TO ORGANIC FARMING PRACTICES, ELIMINATING SYNTHETIC PESTICIDES AND FERTILIZERS.

DAVID HOLMAN

- SOIL REGENERATION: SOIL HEALTH SENSORS PROVIDED REAL-TIME DATA, ENABLING PRECISE NUTRIENT MANAGEMENT AND PROMOTING SOIL REGENERATION.

- PEST CONTROL: A SOLAR-POWERED ELECTRIC FENCE EFFECTIVELY DETERRED PESTS WITHOUT HARMING THE ENVIRONMENT.

LESSONS LEARNED:

- SUSTAINABLE PRACTICES: ADVANCED ELECTROCULTURE SYSTEMS CAN SUPPORT THE TRANSITION TO SUSTAINABLE AND ORGANIC FARMING, REDUCING THE RELIANCE ON SYNTHETIC INPUTS.

- SOIL-CENTRIC APPROACH: CONTINUOUS MONITORING OF SOIL HEALTH IS CRUCIAL FOR OPTIMIZING NUTRIENT

DAVID HOLMAN

MANAGEMENT AND PROMOTING REGENERATIVE AGRICULTURE.

CASE STUDY 4: THE HYDROPONIC ROOFTOP GARDEN

BACKGROUND

LOCATION: NEW YORK CITY, USA

OBJECTIVE: TO ESTABLISH A HYDROPONIC ROOFTOP GARDEN IN AN URBAN ENVIRONMENT AND REDUCE RELIANCE ON GRID ELECTRICITY.

ELECTROCULTURE COMPONENTS:

- EARTH BATTERIES: PARALLEL-CONNECTED EARTH BATTERIES FOR URBAN ENERGY GENERATION.

- CAPACITANCE TECHNOLOGY: ENERGY STORAGE AND DISTRIBUTION.

- SOLAR PANELS: RENEWABLE ENERGY SOURCE.

- DRIP IRRIGATION: EFFICIENT WATER DELIVERY.

KEY ACHIEVEMENTS:

- URBAN FARMING SUCCESS: THE ROOFTOP GARDEN PRODUCED A VARIETY OF VEGETABLES AND HERBS, CONTRIBUTING TO LOCAL FOOD SUSTAINABILITY.

- ENERGY EFFICIENCY: EARTH BATTERIES AND SOLAR PANELS PROVIDED A SELF-SUSTAINING ENERGY SUPPLY FOR LIGHTING, CLIMATE CONTROL, AND IRRIGATION.

- WATER CONSERVATION: DRIP IRRIGATION SYSTEMS MINIMIZED WATER USAGE WHILE MAXIMIZING CROP GROWTH.

LESSONS LEARNED:

- URBAN AGRICULTURE POTENTIAL: ADVANCED ELECTROCULTURE SYSTEMS

CAN EMPOWER URBAN FARMERS TO GROW FRESH PRODUCE SUSTAINABLY IN LIMITED SPACES.

. OFF-GRID FARMING: COMBINING EARTH BATTERIES AND RENEWABLE ENERGY SOURCES ENABLES OFF-GRID FARMING, REDUCING THE ENVIRONMENTAL FOOTPRINT.

CASE STUDY 5: THE RICE PADDY REVOLUTION

BACKGROUND

LOCATION: TAMIL NADU, INDIA

OBJECTIVE: TO IMPROVE RICE CROP YIELDS AND REDUCE WATER USAGE IN A REGION FACING WATER SCARCITY.

ELECTROCULTURE COMPONENTS:

. EARTH BATTERIES: PARALLEL-CONNECTED EARTH BATTERIES FOR ENERGY GENERATION.

- CAPACITANCE TECHNOLOGY: ENERGY STORAGE AND DISTRIBUTION.

- RICE PADDY SENSORS: SOIL MOISTURE AND CROP HEALTH MONITORING.

- DRIP IRRIGATION: WATER-EFFICIENT IRRIGATION.

KEY ACHIEVEMENTS:

- INCREASED YIELDS: PRECISE ELECTRICAL STIMULATION BASED ON SOIL CONDITIONS LED TO A SIGNIFICANT INCREASE IN RICE CROP YIELDS.

- WATER SAVINGS: SOIL MOISTURE SENSORS AND DRIP IRRIGATION REDUCED WATER CONSUMPTION BY OVER 30%.

- SUSTAINABLE FARMING: THE PROJECT SERVED AS A MODEL FOR SUSTAINABLE AND RESOURCE-EFFICIENT RICE

CULTIVATION IN WATER-SCARCE REGIONS.

LESSONS LEARNED:

- **TAILORED CROP MANAGEMENT:** ELECTROCULTURE SYSTEMS CAN FINE-TUNE CROP MANAGEMENT PRACTICES, OPTIMIZING RESOURCE USAGE AND YIELD.

- **WATER-ENERGY NEXUS:** THE SYNERGY OF EARTH BATTERIES, CAPACITANCE TECHNOLOGY, AND ADVANCED SENSORS CAN ADDRESS THE COMPLEX CHALLENGES OF WATER AND ENERGY IN AGRICULTURE.

CASE STUDY 6: THE COMMUNITY FOOD FOREST

BACKGROUND

LOCATION: VANCOUVER, CANADA

OBJECTIVE: TO CREATE A COMMUNITY FOOD FOREST THAT

PROVIDES SUSTAINABLE, LOCALLY GROWN FOOD.

ELECTROCULTURE COMPONENTS:

- EARTH BATTERIES: SERIES-CONNECTED EARTH BATTERIES FOR ENERGY GENERATION.

- CAPACITANCE TECHNOLOGY: ENERGY STORAGE AND DISTRIBUTION.

- ORCHARD SENSORS: CLIMATE AND SOIL MONITORING.

- COMMUNITY ENGAGEMENT: VOLUNTEER AND EDUCATIONAL PROGRAMS.

KEY ACHIEVEMENTS:

- ABUNDANT HARVESTS: THE FOOD FOREST PRODUCED A DIVERSE RANGE OF FRUITS, NUTS, AND VEGETABLES, ENHANCING LOCAL FOOD SECURITY.

DAVID HOLMAN

- ENERGY SELF-RELIANCE: EARTH BATTERIES POWERED ESSENTIAL INFRASTRUCTURE, INCLUDING LIGHTING FOR COMMUNITY EVENTS.

- EDUCATION AND ENGAGEMENT: THE PROJECT ENGAGED THE COMMUNITY IN SUSTAINABLE AGRICULTURE, FOSTERING A SENSE OF OWNERSHIP AND STEWARDSHIP.

LESSONS LEARNED:

- COMMUNITY-LED AGRICULTURE: ELECTROCULTURE SYSTEMS CAN SUPPORT COMMUNITY-LED INITIATIVES, ENHANCING FOOD SECURITY AND SUSTAINABILITY.

- EDUCATION AND INVOLVEMENT: INVOLVING THE COMMUNITY IN ALL ASPECTS OF THE PROJECT, FROM PLANTING TO HARVESTING, FOSTERS A

DEEPER CONNECTION TO THE LAND AND PROMOTES SUSTAINABILITY.

CASE STUDY 7: THE AGTECH INNOVATION HUB

BACKGROUND

LOCATION: SILICON VALLEY, CALIFORNIA, USA

OBJECTIVE: TO DEVELOP AND SHOWCASE CUTTING-EDGE AGRICULTURAL TECHNOLOGY WHILE MINIMIZING ENVIRONMENTAL IMPACT.

ELECTROCULTURE COMPONENTS:

- EARTH BATTERIES: SERIES-CONNECTED EARTH BATTERIES FOR ENERGY GENERATION.

- CAPACITANCE TECHNOLOGY: ENERGY STORAGE AND DISTRIBUTION.

DAVID HOLMAN

- SENSOR NETWORK: ADVANCED SOIL, CLIMATE, AND CROP HEALTH MONITORING.

- RESEARCH AND DEVELOPMENT: COLLABORATION WITH TECH COMPANIES AND UNIVERSITIES.

KEY ACHIEVEMENTS:

- TECHNOLOGICAL INNOVATION: THE HUB SERVED AS A TESTBED FOR EMERGING AGRICULTURAL TECHNOLOGIES, ATTRACTING STARTUPS AND RESEARCHERS.

- SUSTAINABLE PRACTICES: ELECTROCULTURE SYSTEMS ENABLED THE HUB TO MAINTAIN HIGH YIELDS WHILE MINIMIZING RESOURCE USAGE AND ENVIRONMENTAL IMPACT.

- KNOWLEDGE EXCHANGE: COLLABORATION WITH TECH COMPANIES AND UNIVERSITIES RESULTED IN PIONEERING ADVANCEMENTS IN ELECTROCULTURE.

LESSONS LEARNED:

- INNOVATION HUB MODEL: THE INTEGRATION OF ADVANCED ELECTROCULTURE SYSTEMS IN AN INNOVATION HUB CAN DRIVE AGRICULTURAL TECHNOLOGY ADVANCEMENT AND SUSTAINABLE PRACTICES.

- COLLABORATION AND RESEARCH: PARTNERING WITH TECH ORGANIZATIONS AND ACADEMIA CAN LEAD TO GROUNDBREAKING DEVELOPMENTS IN ELECTROCULTURE TECHNOLOGY.

THE LESSONS LEARNED FROM THESE CASE STUDIES EMPHASIZE THE IMPORTANCE OF TAILORED CROP MANAGEMENT, SUSTAINABLE PRACTICES, COMMUNITY INVOLVEMENT, AND INNOVATION IN SHAPING THE FUTURE OF AGRICULTURE. AS WE CONCLUDE

DAVID HOLMAN

THIS CHAPTER, WE ARE REMINDED OF THE VAST POSSIBILITIES THAT ADVANCED ELECTROCULTURE HOLDS FOR ACHIEVING SUSTAINABLE, RESOURCE-EFFICIENT, AND RESILIENT FARMING PRACTICES.

CHAPTER 10

CHALLENGES AND SOLUTIONS

NAVIGATING THE COMPLEXITIES OF ADVANCED ELECTROCULTURE

AS WE VENTURE FURTHER INTO THE WORLD OF ADVANCED ELECTROCULTURE, WE ENCOUNTER NOT ONLY REMARKABLE OPPORTUNITIES BUT ALSO A SPECTRUM OF CHALLENGES. IMPLEMENTING INNOVATIVE TECHNOLOGY IN AGRICULTURE PRESENTS UNIQUE HURDLES, FROM TECHNICAL INTRICACIES TO LOGISTICAL OBSTACLES. IN THIS CHAPTER, WE WILL EXPLORE THE CHALLENGES

FACED BY PRACTITIONERS OF ADVANCED ELECTROCULTURE AND THE INNOVATIVE SOLUTIONS THAT HAVE EMERGED TO ADDRESS THEM.

CHALLENGE 1: ENERGY STORAGE AND DISTRIBUTION

THE CHALLENGE:

- **INTERMITTENT ENERGY PRODUCTION:** EARTH BATTERIES AND RENEWABLE SOURCES SUCH AS SOLAR PANELS PRODUCE ENERGY INTERMITTENTLY, LEADING TO ENERGY SURPLUS AND DEFICIT CYCLES.

- **ENERGY DEMAND VARIATION:** AGRICULTURAL OPERATIONS HAVE FLUCTUATING ENERGY DEMANDS DEPENDING ON FACTORS LIKE CLIMATE CONTROL, IRRIGATION, AND ELECTRICAL STIMULATION.

THE SOLUTION:

- **ENERGY STORAGE:** IMPLEMENT ENERGY STORAGE SOLUTIONS, SUCH AS CAPACITORS OR BATTERIES, TO STORE SURPLUS ENERGY FOR USE DURING PERIODS OF DEFICIT.

- **ADVANCED CONTROL SYSTEMS:** UTILIZE SOPHISTICATED CONTROL SYSTEMS THAT MANAGE ENERGY DISTRIBUTION EFFICIENTLY BASED ON REAL-TIME DEMAND AND AVAILABILITY.

CHALLENGE 2: SOIL VARIABILITY AND CONDUCTIVITY

THE CHALLENGE:

- **SOIL DIVERSITY:** DIFFERENT SOIL TYPES AND CONDITIONS REQUIRE TAILORED APPROACHES FOR OPTIMIZING

CONDUCTIVITY AND ELECTRICAL STIMULATION.

- **SOIL CHANGES OVER TIME:** SOIL COMPOSITION AND CONDUCTIVITY CAN CHANGE DUE TO WEATHER, CROPPING PRACTICES, AND SEASONAL VARIATIONS.

THE SOLUTION:

- **SOIL SENSORS:** DEPLOY SOIL SENSORS THAT CONTINUOUSLY MONITOR SOIL MOISTURE, TEMPERATURE, AND NUTRIENT LEVELS, ALLOWING FOR REAL-TIME ADJUSTMENTS TO ELECTROCULTURE PARAMETERS.

- **VARIABLE ELECTRICAL STIMULATION:** DEVELOP ALGORITHMS THAT ADAPT ELECTRICAL STIMULATION BASED ON SOIL DATA, ENSURING OPTIMAL

CONDITIONS FOR PLANT GROWTH.

CHALLENGE 3: TECHNICAL EXPERTISE

THE CHALLENGE:

- **COMPLEX TECHNOLOGY:** ADVANCED ELECTROCULTURE SYSTEMS INVOLVE INTRICATE ELECTRICAL COMPONENTS, SENSORS, AND CONTROL SYSTEMS.

- **TRAINING AND KNOWLEDGE GAPS:** MANY FARMERS LACK THE TECHNICAL EXPERTISE NEEDED TO DESIGN, INSTALL, AND MAINTAIN THESE SYSTEMS.

THE SOLUTION:

- **TRAINING PROGRAMS:** ESTABLISH TRAINING PROGRAMS AND WORKSHOPS TO EDUCATE FARMERS AND AGRICULTURAL PROFESSIONALS ON ADVANCED

ELECTROCULTURE TECHNOLOGY.

. **CONSULTING SERVICES:** OFFER CONSULTING SERVICES PROVIDED BY EXPERTS IN THE FIELD TO ASSIST WITH SYSTEM DESIGN, INSTALLATION, AND TROUBLESHOOTING.

CHALLENGE 4: COST OF IMPLEMENTATION

THE CHALLENGE:

. **INITIAL INVESTMENT:** ADVANCED ELECTROCULTURE SYSTEMS REQUIRE A SIGNIFICANT UPFRONT INVESTMENT IN EQUIPMENT, SENSORS, AND TECHNOLOGY.

. **ROI UNCERTAINTY:** FARMERS MAY BE UNCERTAIN ABOUT THE RETURN ON INVESTMENT (ROI) AND LONG-TERM COST SAVINGS.

THE SOLUTION:

- **FINANCING OPTIONS:** EXPLORE FINANCING OPTIONS AND INCENTIVES FOR ADOPTING ADVANCED ELECTROCULTURE, SUCH AS GRANTS, SUBSIDIES, OR LOW-INTEREST LOANS.

- **ROI DEMONSTRATIONS:** SHARE CASE STUDIES AND ROI DATA FROM SUCCESSFUL IMPLEMENTATIONS TO ILLUSTRATE THE FINANCIAL BENEFITS OVER TIME.

CHALLENGE 5: ENVIRONMENTAL IMPACT

THE CHALLENGE:

- **SUSTAINABILITY CONCERNS:** THE MANUFACTURING AND DISPOSAL OF ELECTRICAL COMPONENTS CAN HAVE ENVIRONMENTAL IMPACTS, POTENTIALLY COUNTERACTING

THE SUSTAINABILITY GOALS OF ADVANCED ELECTROCULTURE.

- **ENERGY SOURCE CONSIDERATIONS:** THE SOURCE OF ELECTRICITY FOR CHARGING EARTH BATTERIES CAN AFFECT THE OVERALL ENVIRONMENTAL FOOTPRINT.

THE SOLUTION:

- **ECO-FRIENDLY COMPONENTS:** ENCOURAGE THE USE OF ECO-FRIENDLY MATERIALS AND TECHNOLOGIES IN THE CONSTRUCTION OF ELECTRICAL COMPONENTS.

- **RENEWABLE ENERGY INTEGRATION:** PRIORITIZE RENEWABLE ENERGY SOURCES, SUCH AS SOLAR OR WIND, FOR CHARGING EARTH BATTERIES TO REDUCE CARBON EMISSIONS.

CHALLENGE 6: PEST AND WEED MANAGEMENT

THE CHALLENGE:

- **ELECTROCULTURE EFFECTS ON PESTS AND WEEDS:** THE APPLICATION OF ELECTRICAL STIMULATION MAY INADVERTENTLY AFFECT PEST POPULATIONS OR WEED GROWTH.

- **INTEGRATED PEST MANAGEMENT (IPM):** INTEGRATING ELECTROCULTURE WITH TRADITIONAL IPM PRACTICES CAN BE COMPLEX.

THE SOLUTION:

- **MONITORING AND CONTROL:** IMPLEMENT MONITORING SYSTEMS THAT TRACK PEST AND WEED POPULATIONS, ALLOWING FOR EARLY INTERVENTION IF NECESSARY.

- **ADAPTIVE IPM:** DEVELOP ADAPTIVE IPM STRATEGIES

THAT INCORPORATE ELECTROCULTURE PRACTICES WHILE ADDRESSING PEST AND WEED CHALLENGES.

CHALLENGE 7: REGULATORY AND COMPLIANCE ISSUES

THE CHALLENGE:

- **LACK OF STANDARDS:** THE FIELD OF ADVANCED ELECTROCULTURE LACKS STANDARDIZED PRACTICES AND REGULATIONS, MAKING IT CHALLENGING FOR FARMERS TO NAVIGATE.

- **PERMITTING AND SAFETY:** INSTALLING ELECTRICAL SYSTEMS ON AGRICULTURAL LAND MAY REQUIRE PERMITS AND COMPLIANCE WITH SAFETY STANDARDS.

THE SOLUTION:

- **INDUSTRY COLLABORATION:**

COLLABORATE WITH INDUSTRY STAKEHOLDERS TO DEVELOP GUIDELINES AND BEST PRACTICES FOR ADVANCED ELECTROCULTURE.

. **CONSULTATION WITH AUTHORITIES:** WORK CLOSELY WITH REGULATORY AUTHORITIES AND SAFETY AGENCIES TO ENSURE COMPLIANCE AND SAFETY IN SYSTEM INSTALLATIONS.

CHALLENGE 8: ADAPTATION TO CLIMATE CHANGE

THE CHALLENGE:

. **CLIMATE UNCERTAINTY:** CHANGING CLIMATE PATTERNS AND EXTREME WEATHER EVENTS CAN DISRUPT ELECTROCULTURE OPERATIONS.

. **WATER SCARCITY:** INCREASED WATER SCARCITY

IN SOME REGIONS CHALLENGES
IRRIGATION PRACTICES.

THE SOLUTION:

- **CLIMATE-RESILIENT SYSTEMS:** DESIGN ELECTROCULTURE SYSTEMS THAT CAN WITHSTAND AND ADAPT TO CLIMATE VARIATIONS, SUCH AS EXCESS RAIN OR PROLONGED DROUGHT.

- **WATER-EFFICIENT TECHNOLOGIES:** EXPLORE ADVANCED IRRIGATION TECHNOLOGIES, INCLUDING PRECISION DRIP SYSTEMS, TO OPTIMIZE WATER USAGE IN ELECTROCULTURE.

CHALLENGE 9: DATA MANAGEMENT AND SECURITY

THE CHALLENGE:

- **DATA OVERLOAD:** THE PROLIFERATION OF SENSORS

GENERATES VAST AMOUNTS OF DATA, WHICH CAN BE OVERWHELMING WITHOUT PROPER MANAGEMENT.

. **DATA SECURITY:** PROTECTING SENSITIVE AGRICULTURAL DATA FROM CYBER THREATS IS A GROWING CONCERN.

THE SOLUTION:

. **DATA ANALYTICS:** UTILIZE DATA ANALYTICS TOOLS TO PROCESS AND EXTRACT MEANINGFUL INSIGHTS FROM THE COLLECTED DATA, AIDING IN DECISION-MAKING.

. **CYBERSECURITY MEASURES:** IMPLEMENT ROBUST CYBERSECURITY MEASURES, INCLUDING ENCRYPTION AND SECURE DATA STORAGE, TO PROTECT SENSITIVE DATA.

DAVID HOLMAN

CHALLENGE 10: SOCIAL ACCEPTANCE AND PERCEPTION

THE CHALLENGE:

- **CONSUMER PERCEPTION:** SOME CONSUMERS MAY BE SKEPTICAL OR UNAWARE OF ELECTROCULTURE PRACTICES, IMPACTING THE MARKET FOR PRODUCTS GROWN WITH THESE METHODS.

- **CULTURAL AND SOCIAL FACTORS:** ELECTROCULTURE MAY FACE RESISTANCE OR SKEPTICISM IN CERTAIN REGIONS DUE TO CULTURAL OR SOCIAL FACTORS.

THE SOLUTION:

- **EDUCATION AND OUTREACH:** LAUNCH EDUCATIONAL CAMPAIGNS TO INFORM CONSUMERS ABOUT THE BENEFITS AND SAFETY OF ELECTROCULTURE-GROWN PRODUCTS.

DAVID HOLMAN

- **COMMUNITY ENGAGEMENT:** ENGAGE WITH LOCAL COMMUNITIES TO ADDRESS CONCERNS AND FOSTER UNDERSTANDING OF ELECTROCULTURE PRACTICES.

CHALLENGE 11: LONG-TERM SUSTAINABILITY

THE CHALLENGE:

- **RESOURCE DEPLETION:** PROLONGED AND INTENSIVE ELECTROCULTURE PRACTICES MAY DEPLETE SOIL NUTRIENTS AND DISRUPT LONG-TERM SOIL HEALTH.

- **ENERGY SOURCE RELIABILITY:** THE RELIABILITY OF RENEWABLE ENERGY SOURCES FOR CHARGING EARTH BATTERIES OVER DECADES IS UNCERTAIN.

THE SOLUTION:

- **CROP ROTATION AND COVER CROPS:** IMPLEMENT CROP ROTATION AND COVER CROPPING TO REPLENISH SOIL NUTRIENTS AND MAINTAIN SOIL HEALTH.

- **DIVERSIFIED ENERGY SOURCES:** EXPLORE A DIVERSIFIED ENERGY STRATEGY THAT INCLUDES BACKUP ENERGY SOURCES FOR PROLONGED SUSTAINABILITY.

CHALLENGE 12: SCALING AND INTEGRATION

THE CHALLENGE:

- **SCALING COMPLEXITY:** SCALING ADVANCED ELECTROCULTURE SYSTEMS FROM SMALL TRIALS TO LARGE COMMERCIAL OPERATIONS CAN BE COMPLEX.

- **INTEGRATION WITH EXISTING SYSTEMS:** INTEGRATING

ELECTROCULTURE PRACTICES INTO EXISTING FARMING OPERATIONS CAN DISRUPT ESTABLISHED ROUTINES.

THE SOLUTION:

- **PILOT PROJECTS:** CONDUCT PILOT PROJECTS TO TEST SCALABILITY AND IDENTIFY CHALLENGES BEFORE FULL-SCALE IMPLEMENTATION.

- **GRADUAL INTEGRATION:** GRADUALLY INTEGRATE ELECTROCULTURE PRACTICES INTO EXISTING FARMING OPERATIONS, ALLOWING FOR ADJUSTMENT AND ADAPTATION.

CHALLENGE 13: COLLABORATION AND KNOWLEDGE SHARING

THE CHALLENGE:

- **ISOLATION AND LACK OF INFORMATION:** ISOLATED OR REMOTE FARMERS MAY LACK

ACCESS TO INFORMATION, RESOURCES, AND PEER NETWORKS FOR ADOPTING ADVANCED ELECTROCULTURE.

- **INTELLECTUAL PROPERTY CONCERNS:** FEAR OF INTELLECTUAL PROPERTY RIGHTS ISSUES MAY HINDER COLLABORATION AND KNOWLEDGE SHARING.

THE SOLUTION:

- **ONLINE RESOURCES:** ESTABLISH ONLINE PLATFORMS AND FORUMS FOR KNOWLEDGE SHARING, WHERE FARMERS CAN ACCESS INFORMATION AND COLLABORATE.

- **OPEN-SOURCE INITIATIVES:** PROMOTE OPEN-SOURCE SOLUTIONS AND COLLABORATIVE PROJECTS THAT FACILITATE INFORMATION EXCHANGE WHILE RESPECTING

INTELLECTUAL PROPERTY RIGHTS.

CHALLENGE 14: TECHNOLOGICAL EVOLUTION

THE CHALLENGE:

- **RAPID TECHNOLOGICAL CHANGE:** THE FAST-PACED EVOLUTION OF TECHNOLOGY MAY MAKE CURRENT ELECTROCULTURE SYSTEMS OUTDATED QUICKLY.

- **COMPATIBILITY ISSUES:** INTEGRATING NEW TECHNOLOGY INTO EXISTING SYSTEMS CAN POSE COMPATIBILITY CHALLENGES.

THE SOLUTION:

- **CONTINUOUS LEARNING:** STAY UPDATED ON EMERGING TECHNOLOGIES AND CONSIDER SCALABILITY AND ADAPTABILITY WHEN

DESIGNING ELECTROCULTURE SYSTEMS.

. **MODULAR SYSTEMS:** BUILD MODULAR ELECTROCULTURE SYSTEMS THAT ALLOW FOR THE EASY INTEGRATION OF NEW TECHNOLOGY COMPONENTS.

THE CHALLENGES ENCOUNTERED IN THE REALM OF ADVANCED ELECTROCULTURE ARE MULTIFACETED AND DEMAND INNOVATIVE SOLUTIONS. AS WE HAVE EXPLORED IN THIS CHAPTER, ADDRESSING ISSUES RELATED TO ENERGY STORAGE, SOIL VARIABILITY, TECHNICAL EXPERTISE, AND COST IS CRUCIAL FOR THE WIDESPREAD ADOPTION OF ELECTROCULTURE PRACTICES.

FURTHERMORE, THE NEED FOR ENVIRONMENTAL SUSTAINABILITY, PEST AND WEED MANAGEMENT, COMPLIANCE WITH

REGULATIONS, AND ADAPTATION
TO CLIMATE CHANGE
NECESSITATES CREATIVE
PROBLEM-SOLVING. DATA
MANAGEMENT, SOCIAL
ACCEPTANCE, LONG-TERM
SUSTAINABILITY, SCALING,
COLLABORATION, AND
TECHNOLOGICAL EVOLUTION ARE
ALSO CRITICAL AREAS THAT
REQUIRE ONGOING ATTENTION
AND INNOVATION.

CHAPTER 11

FUTURE DEVELOPMENTS AND TRENDS

EMERGING TECHNOLOGIES AND THE FUTURE OF ADVANCED ELECTROCULTURE

AS WE EMBARK ON THE FINAL
CHAPTER OF OUR EXPLORATION
INTO ADVANCED
ELECTROCULTURE, WE CAST OUR
GAZE TOWARD THE FUTURE. THE

FIELD OF ELECTROCULTURE, MARKED BY ITS INNOVATIVE INTEGRATION OF EARTH BATTERIES, CAPACITANCE TECHNOLOGY, AND SUSTAINABLE AGRICULTURE PRACTICES, CONTINUES TO EVOLVE RAPIDLY. IN THIS CHAPTER, WE WILL SPECULATE ON EMERGING TECHNOLOGIES, TRENDS, AND THE POTENTIAL IMPACT OF ELECTROCULTURE ON GLOBAL AGRICULTURE IN THE YEARS TO COME.

SECTION 1: THE PROMISE OF ADVANCEMENTS

1.1 QUANTUM-DOT ENERGY STORAGE

CONCEPT: QUANTUM DOTS, NANOSCALE SEMICONDUCTOR PARTICLES, SHOW PROMISE IN REVOLUTIONIZING ENERGY STORAGE IN ELECTROCULTURE SYSTEMS. THEIR UNIQUE

PROPERTIES, SUCH AS TUNABLE BANDGAPS AND HIGH CHARGE STORAGE CAPACITIES, COULD LEAD TO ULTRA-EFFICIENT AND COMPACT ENERGY STORAGE SOLUTIONS.

POTENTIAL IMPACT: QUANTUM-DOT-BASED ENERGY STORAGE SYSTEMS COULD SIGNIFICANTLY INCREASE THE ENERGY DENSITY OF EARTH BATTERIES, REDUCING THE FOOTPRINT OF ENERGY STORAGE INFRASTRUCTURE.

CHALLENGES: CHALLENGES INCLUDE SCALABILITY, COST, AND LONG-TERM STABILITY OF QUANTUM-DOT-BASED STORAGE SYSTEMS.

1.2 BIOLOGICALLY ENHANCED ELECTRODES

CONCEPT: HARNESSING THE POWER OF MICROORGANISMS TO ENHANCE EARTH BATTERY ELECTRODES. BY PROMOTING

BENEFICIAL MICROBIAL ACTIVITY, THESE ELECTRODES CAN FACILITATE IMPROVED ELECTRON TRANSFER AND ENERGY PRODUCTION.

POTENTIAL IMPACT: BIOLOGICALLY ENHANCED ELECTRODES COULD BOOST THE EFFICIENCY OF EARTH BATTERIES, MAKING THEM MORE RELIABLE AND SUSTAINABLE.

CHALLENGES: DEVELOPING SUITABLE ELECTRODE MATERIALS, MAINTAINING MICROBIAL BALANCE, AND AVOIDING DETRIMENTAL MICROBIAL ACTIVITY ARE KEY CHALLENGES.

SECTION 2: EXPANDING APPLICATIONS

2.1 VERTICAL FARMING INTEGRATION

CONCEPT: THE INTEGRATION OF ELECTROCULTURE INTO VERTICAL FARMING SYSTEMS. BY

OPTIMIZING ENERGY USAGE AND PLANT STIMULATION, VERTICAL FARMS CAN ACHIEVE HIGHER YIELDS WITH REDUCED RESOURCE CONSUMPTION.

POTENTIAL IMPACT: VERTICAL FARMING OFFERS THE POTENTIAL FOR YEAR-ROUND, CLIMATE-INDEPENDENT AGRICULTURE IN URBAN AREAS, HELPING ADDRESS FOOD SECURITY CHALLENGES.

CHALLENGES: DESIGNING ELECTROCULTURE SYSTEMS THAT FIT THE VERTICAL FARMING INFRASTRUCTURE AND DEVELOPING EFFICIENT LIGHTING AND CLIMATE CONTROL SOLUTIONS.

2.2 SMART AGRICULTURAL GRIDS

CONCEPT: THE CREATION OF INTERCONNECTED, SMART AGRICULTURAL GRIDS THAT ENABLE THE EXCHANGE OF

SURPLUS ENERGY BETWEEN
FARMS. THIS GRID CAN BALANCE
ENERGY SUPPLY AND DEMAND,
ENHANCING ENERGY EFFICIENCY.

POTENTIAL IMPACT: SMART
GRIDS CAN HELP RURAL
COMMUNITIES BECOME ENERGY
SELF-SUFFICIENT, REDUCING
THEIR RELIANCE ON CENTRALIZED
ENERGY SOURCES.

CHALLENGES: ESTABLISHING
THE INFRASTRUCTURE AND
STANDARDS FOR SMART
AGRICULTURAL GRIDS AND
ADDRESSING GRID SECURITY
CONCERNS.

SECTION 3: SUSTAINABILITY AND CONSERVATION

3.1 ELECTROCULTURE FOR DESERT RECLAMATION

CONCEPT: THE APPLICATION OF
ELECTROCULTURE IN DESERT
RECLAMATION PROJECTS.
ELECTROCULTURE CAN OPTIMIZE

SOIL CONDUCTIVITY AND MOISTURE LEVELS, MAKING DESERT LAND ARABLE.

POTENTIAL IMPACT: DESERT RECLAMATION CAN EXPAND AGRICULTURAL LAND AND IMPROVE FOOD SECURITY WHILE COMBATING DESERTIFICATION.

CHALLENGES: ADAPTING ELECTROCULTURE SYSTEMS TO EXTREME DESERT CONDITIONS, WATER SOURCING, AND COST-EFFECTIVENESS.

3.2 ELECTROCULTURE FOR SOIL CARBON SEQUESTRATION

CONCEPT: USING ELECTROCULTURE PRACTICES TO ENHANCE SOIL CARBON SEQUESTRATION. CONTROLLED ELECTRICAL STIMULATION CAN PROMOTE MICROBIAL ACTIVITY AND INCREASE ORGANIC MATTER IN THE SOIL.

POTENTIAL IMPACT: SOIL CARBON SEQUESTRATION CAN MITIGATE CLIMATE CHANGE BY CAPTURING CARBON DIOXIDE FROM THE ATMOSPHERE AND IMPROVING SOIL FERTILITY.

CHALLENGES: DEVELOPING PRECISE STIMULATION PROTOCOLS AND MONITORING SYSTEMS FOR CARBON SEQUESTRATION.

SECTION 4: BEYOND EARTH BATTERIES

4.1 ADVANCES IN ENERGY HARVESTING

CONCEPT: THE DEVELOPMENT OF INNOVATIVE ENERGY HARVESTING TECHNOLOGIES, SUCH AS TRIBOELECTRIC GENERATORS AND PIEZOELECTRIC MATERIALS, TO COMPLEMENT OR REPLACE EARTH BATTERIES.

POTENTIAL IMPACT: THESE TECHNOLOGIES CAN GENERATE

ENERGY FROM ENVIRONMENTAL SOURCES LIKE MECHANICAL VIBRATIONS OR FRICTION, EXPANDING THE ENERGY SOURCES AVAILABLE FOR ELECTROCULTURE.

CHALLENGES: EFFICIENCY AND SCALABILITY OF ENERGY HARVESTING TECHNOLOGIES.

4.2 INTEGRATION WITH ADVANCED AI AND MACHINE LEARNING

CONCEPT: THE INTEGRATION OF ADVANCED ARTIFICIAL INTELLIGENCE (AI) AND MACHINE LEARNING (ML) ALGORITHMS INTO ELECTROCULTURE SYSTEMS. AI CAN OPTIMIZE ENERGY DISTRIBUTION, PEST CONTROL, AND CROP MANAGEMENT BASED ON VAST AMOUNTS OF DATA.

POTENTIAL IMPACT: AI-DRIVEN ELECTROCULTURE SYSTEMS CAN ENHANCE PRECISION FARMING,

RESOURCE OPTIMIZATION, AND OVERALL SUSTAINABILITY.

CHALLENGES: DATA PRIVACY CONCERNS, DEVELOPMENT OF ROBUST AI MODELS, AND ACCESSIBILITY TO AI TECHNOLOGY.

SECTION 5: GLOBAL IMPACT AND CHALLENGES

5.1 ELECTROCULTURE IN DEVELOPING REGIONS

CONCEPT: THE ADOPTION OF ELECTROCULTURE PRACTICES IN DEVELOPING REGIONS TO ADDRESS FOOD SECURITY CHALLENGES. LOW-COST, SCALABLE ELECTROCULTURE SOLUTIONS CAN EMPOWER SMALLHOLDER FARMERS.

POTENTIAL IMPACT: ELECTROCULTURE CAN IMPROVE CROP YIELDS, REDUCE RESOURCE USAGE, AND ALLEVIATE POVERTY IN REGIONS WITH LIMITED ACCESS

TO MODERN AGRICULTURAL TECHNOLOGY.

CHALLENGES: ACCESSIBILITY TO TECHNOLOGY, INFRASTRUCTURE DEVELOPMENT, AND KNOWLEDGE DISSEMINATION.

5.2 ETHICAL CONSIDERATIONS

CONCEPT: THE ETHICAL IMPLICATIONS OF ADVANCED ELECTROCULTURE, INCLUDING QUESTIONS ABOUT ENVIRONMENTAL IMPACT, LONG-TERM SUSTAINABILITY, AND SOCIAL ACCEPTANCE.

POTENTIAL IMPACT: ETHICAL CONSIDERATIONS CAN SHAPE THE DEVELOPMENT AND ADOPTION OF ELECTROCULTURE PRACTICES, ENSURING THEY ALIGN WITH SOCIETAL VALUES AND GOALS.

CHALLENGES: BALANCING TECHNOLOGICAL ADVANCEMENTS WITH ETHICAL CONCERNS AND

ADDRESSING ETHICAL DILEMMAS THAT MAY ARISE.

SECTION 6: A VISION FOR THE FUTURE

6.1 THE GLOBAL ELECTROCULTURE NETWORK

CONCEPT: THE ESTABLISHMENT OF A GLOBAL NETWORK OF ELECTROCULTURE PRACTITIONERS, RESEARCHERS, AND ORGANIZATIONS. THIS NETWORK FOSTERS COLLABORATION, KNOWLEDGE EXCHANGE, AND THE DEVELOPMENT OF STANDARDIZED PRACTICES.

POTENTIAL IMPACT: THE GLOBAL ELECTROCULTURE NETWORK CAN ACCELERATE INNOVATION AND THE ADOPTION OF ELECTROCULTURE PRACTICES WORLDWIDE.

CHALLENGES: COORDINATING DIVERSE STAKEHOLDERS AND

MANAGING A GLOBAL NETWORK EFFECTIVELY.

6.2 THE NEXT AGRICULTURAL REVOLUTION

CONCEPT: ELECTROCULTURE AS THE CATALYST FOR THE NEXT AGRICULTURAL REVOLUTION, USHERING IN AN ERA OF SUSTAINABLE, EFFICIENT, AND TECHNOLOGICALLY ADVANCED FARMING PRACTICES.

POTENTIAL IMPACT: THE WIDESPREAD ADOPTION OF ELECTROCULTURE CAN ADDRESS FOOD SECURITY, RESOURCE CONSERVATION, AND CLIMATE CHANGE MITIGATION ON A GLOBAL SCALE.

CHALLENGES: OVERCOMING THE BARRIERS TO ADOPTION, INCLUDING RESISTANCE TO CHANGE AND THE CHALLENGES OUTLINED IN THIS BOOK.

AS WE CONCLUDE OUR JOURNEY
THROUGH THE REALM OF
ADVANCED ELECTROCULTURE, WE
FIND OURSELVES STANDING AT
THE THRESHOLD OF AN EXCITING
AND TRANSFORMATIVE FUTURE.
EMERGING TECHNOLOGIES,
EXPANDING APPLICATIONS,
SUSTAINABILITY INITIATIVES,
AND ETHICAL CONSIDERATIONS
ARE SHAPING THE TRAJECTORY
OF ELECTROCULTURE. THE
POTENTIAL IMPACT OF
ELECTROCULTURE EXTENDS
BEYOND FARMING FIELDS,
TOUCHING ON GLOBAL FOOD
SECURITY, ENVIRONMENTAL
CONSERVATION, AND
TECHNOLOGICAL INNOVATION.

THE FUTURE OF
ELECTROCULTURE HINGES ON
COLLABORATIVE EFFORTS,
KNOWLEDGE SHARING, AND A
COMMITMENT TO
SUSTAINABILITY. IT REPRESENTS

DAVID HOLMAN

NOT ONLY A PROMISING AVENUE FOR AGRICULTURAL ADVANCEMENT BUT ALSO A TESTAMENT TO HUMAN INGENUITY AND OUR CAPACITY TO HARMONIZE TECHNOLOGY WITH NATURE.

AS WE LOOK AHEAD, LET US ENVISION A WORLD WHERE ADVANCED ELECTROCULTURE PLAYS A PIVOTAL ROLE IN NOURISHING OUR PLANET SUSTAINABLY AND SECURING A BRIGHTER FUTURE FOR GENERATIONS TO COME.

CHAPTER 12

CONCLUSION

KEY TAKEAWAYS, ENCOURAGEMENT, AND THE PATH TO SUSTAINABLE AGRICULTURE

AS WE DRAW THE FINAL CURTAIN ON OUR JOURNEY THROUGH THE

WORLD OF ADVANCED ELECTROCULTURE, IT IS A MOMENT FOR REFLECTION, SYNTHESIS, AND INSPIRATION. THIS CHAPTER SERVES AS A CULMINATION OF OUR EXPLORATION, BRINGING TOGETHER KEY TAKEAWAYS, WORDS OF ENCOURAGEMENT, AND A VISION FOR THE PATH TO SUSTAINABLE AGRICULTURE THROUGH THE LENS OF ELECTROCULTURE.

SECTION 1: RECAPITULATION OF KEY TAKEAWAYS

1.1 ELECTROCULTURE AS A SUSTAINABLE PARADIGM

KEY TAKEAWAY: ELECTROCULTURE REPRESENTS A SUSTAINABLE PARADIGM FOR AGRICULTURE, WHERE THE HARMONIOUS INTEGRATION OF TECHNOLOGY, NATURE, AND HUMAN INGENUITY FOSTERS

RESOURCE-EFFICIENT AND
RESILIENT FARMING PRACTICES.

IMPLICATIONS: BY ADOPTING
ELECTROCULTURE PRINCIPLES,
FARMERS AND AGRICULTURAL
STAKEHOLDERS CAN ADDRESS
CRITICAL CHALLENGES, SUCH AS
RESOURCE DEPLETION, CLIMATE
CHANGE, AND FOOD SECURITY,
WHILE PROMOTING
ENVIRONMENTAL STEWARDSHIP.

1.2 THE POWER OF EARTH BATTERIES

KEY TAKEAWAY: EARTH
BATTERIES, HARNESSED FROM
THE EARTH'S NATURAL ENERGY,
SERVE AS A FOUNDATIONAL
ELEMENT OF ADVANCED
ELECTROCULTURE, PROVIDING A
RELIABLE AND SUSTAINABLE
SOURCE OF ELECTRICAL POWER.

IMPLICATIONS: EARTH
BATTERIES EMPOWER FARMERS
TO REDUCE RELIANCE ON

TRADITIONAL ENERGY SOURCES, MINIMIZE CARBON FOOTPRINTS, AND OPTIMIZE ENERGY UTILIZATION IN AGRICULTURAL OPERATIONS.

1.3 CAPACITANCE TECHNOLOGY'S ROLE

KEY TAKEAWAY: CAPACITANCE TECHNOLOGY ENHANCES ELECTROCULTURE SYSTEMS BY PROVIDING ENERGY STORAGE, EFFICIENT DISTRIBUTION, AND CONTROLLED RELEASE OF ELECTRICAL ENERGY TO STIMULATE PLANT GROWTH.

IMPLICATIONS: CAPACITANCE TECHNOLOGY ENABLES PRECISE AND RESPONSIVE MANAGEMENT OF ELECTRICAL STIMULATION, CONTRIBUTING TO IMPROVED CROP YIELDS, WATER CONSERVATION, AND ENERGY EFFICIENCY.

1.4 INTEGRATION FOR SYNERGY

KEY TAKEAWAY: THE INTEGRATION OF EARTH BATTERIES AND CAPACITANCE TECHNOLOGY, COMBINED WITH ADVANCED SENSORS AND CONTROL SYSTEMS, CREATES SYNERGISTIC EFFECTS THAT DRIVE SUSTAINABLE AND OPTIMIZED AGRICULTURE.

IMPLICATIONS: INTEGRATED ELECTROCULTURE SYSTEMS ENABLE FARMERS TO MAKE DATA-DRIVEN DECISIONS, CONSERVE RESOURCES, AND MAINTAIN CROP HEALTH, LEADING TO INCREASED AGRICULTURAL PRODUCTIVITY AND PROFITABILITY.

1.5 CHALLENGES AND SOLUTIONS

KEY TAKEAWAY: THE JOURNEY TOWARDS ADVANCED ELECTROCULTURE IS NOT

WITHOUT CHALLENGES, BUT INNOVATIVE SOLUTIONS ARE EMERGING TO ADDRESS ISSUES RELATED TO ENERGY STORAGE, SOIL VARIABILITY, TECHNICAL EXPERTISE, AND MORE.

IMPLICATIONS: OVERCOMING CHALLENGES IS AN INTEGRAL PART OF ADVANCING ELECTROCULTURE, AND PRACTITIONERS ARE FINDING CREATIVE SOLUTIONS TO TECHNICAL, FINANCIAL, AND ENVIRONMENTAL OBSTACLES.

SECTION 2: WORDS OF ENCOURAGEMENT

2.1 EMBRACING INNOVATION

ENCOURAGEMENT: EMBRACE INNOVATION WITH AN OPEN MIND AND A WILLINGNESS TO ADAPT. ADVANCED ELECTROCULTURE REPRESENTS A TRANSFORMATIVE APPROACH TO AGRICULTURE, AND THOSE WHO ARE OPEN TO CHANGE

STAND TO REAP ITS MANY REWARDS.

GUIDANCE: START SMALL, EXPERIMENT, AND LEARN FROM YOUR EXPERIENCES. INNOVATION OFTEN BEGINS AT THE GRASSROOTS LEVEL, AND YOUR JOURNEY IN ELECTROCULTURE CAN BE A CATALYST FOR POSITIVE CHANGE IN YOUR COMMUNITY AND BEYOND.

2.2 SUSTAINABLE STEWARDSHIP

ENCOURAGEMENT: EMBRACE YOUR ROLE AS A STEWARD OF THE LAND AND A CUSTODIAN OF THE ENVIRONMENT. ELECTROCULTURE EMPOWERS YOU TO PRACTICE SUSTAINABLE AGRICULTURE WHILE NURTURING THE EARTH FOR FUTURE GENERATIONS.

GUIDANCE: PRIORITIZE REGENERATIVE FARMING PRACTICES THAT RESTORE SOIL

HEALTH, CONSERVE WATER, AND REDUCE ENVIRONMENTAL IMPACT. AS AN ELECTROCULTURE PRACTITIONER, YOU PLAY A VITAL ROLE IN PRESERVING THE PLANET'S RESOURCES.

2.3 COLLABORATION AND KNOWLEDGE SHARING

ENCOURAGEMENT: COLLABORATE WITH FELLOW ELECTROCULTURE ENTHUSIASTS, RESEARCHERS, AND ORGANIZATIONS. THE EXCHANGE OF KNOWLEDGE AND EXPERIENCES ACCELERATES THE ADOPTION OF ADVANCED ELECTROCULTURE AND FUELS COLLECTIVE PROGRESS.

GUIDANCE: JOIN OR ESTABLISH LOCAL AND GLOBAL NETWORKS OF ELECTROCULTURE PRACTITIONERS. SHARE YOUR INSIGHTS, SUCCESSES, AND

CHALLENGES. TOGETHER, WE CAN SHAPE THE FUTURE OF FARMING.

2.4 LIFELONG LEARNING

ENCOURAGEMENT: COMMIT TO LIFELONG LEARNING AND STAY INFORMED ABOUT EMERGING TECHNOLOGIES AND BEST PRACTICES IN ELECTROCULTURE. THE FIELD IS EVOLVING, AND CONTINUOUS EDUCATION IS KEY TO STAYING AT THE FOREFRONT.

GUIDANCE: ENGAGE IN WORKSHOPS, SEMINARS, AND ONLINE COMMUNITIES RELATED TO ELECTROCULTURE. BE CURIOUS, ASK QUESTIONS, AND SEEK OUT MENTORS WHO CAN GUIDE YOUR JOURNEY.

2.5 A VISION FOR THE FUTURE

ENCOURAGEMENT: ENVISION A FUTURE WHERE ELECTROCULTURE IS A CORNERSTONE OF SUSTAINABLE AGRICULTURE WORLDWIDE.

PICTURE THRIVING FARMS, ABUNDANT HARVESTS, AND A HEALTHIER PLANET THANKS TO YOUR CONTRIBUTIONS.

GUIDANCE: KEEP YOUR VISION ALIVE BY SETTING CLEAR GOALS AND MILESTONES. WORK TIRELESSLY TOWARD A FUTURE WHERE ELECTROCULTURE PRINCIPLES ARE EMBRACED BY FARMERS, POLICYMAKERS, AND COMMUNITIES ALIKE.

SECTION 3: THE PATH TO SUSTAINABLE AGRICULTURE

3.1 EDUCATION AND ADVOCACY

PATHWAY: EDUCATION IS THE FOUNDATION OF CHANGE. ADVOCATE FOR THE INCLUSION OF ELECTROCULTURE PRINCIPLES IN AGRICULTURAL CURRICULA AND PROMOTE AWARENESS AMONG POLICYMAKERS, FARMERS, AND THE PUBLIC.

ACTION STEPS:

- COLLABORATE WITH EDUCATIONAL INSTITUTIONS TO DEVELOP ELECTROCULTURE COURSES AND TRAINING PROGRAMS.

- ENGAGE WITH LOCAL AGRICULTURAL EXTENSION SERVICES TO INTRODUCE ELECTROCULTURE TO FARMERS.

- PARTICIPATE IN CONFERENCES, SEMINARS, AND OUTREACH EVENTS TO SHARE YOUR ELECTROCULTURE KNOWLEDGE.

3.2 RESEARCH AND DEVELOPMENT

PATHWAY: RESEARCH IS THE DRIVING FORCE BEHIND INNOVATION. SUPPORT AND ENGAGE IN RESEARCH PROJECTS THAT EXPLORE ELECTROCULTURE'S POTENTIAL

AND ITS ADAPTATION TO DIVERSE AGRICULTURAL CONTEXTS.

ACTION STEPS:

- ESTABLISH PARTNERSHIPS WITH RESEARCH INSTITUTIONS TO CONDUCT ELECTROCULTURE EXPERIMENTS AND STUDIES.

- ALLOCATE RESOURCES TO FUND RESEARCH INITIATIVES AND TECHNOLOGY DEVELOPMENT IN ELECTROCULTURE.

- ENCOURAGE INTERDISCIPLINARY COLLABORATION BETWEEN AGRICULTURAL SCIENCE, ENGINEERING, AND ENVIRONMENTAL STUDIES.

3.3 SUSTAINABLE AGRICULTURE POLICIES

PATHWAY: ADVOCATE FOR POLICIES THAT SUPPORT SUSTAINABLE AGRICULTURAL PRACTICES, INCLUDING

ELECTROCULTURE. ENGAGE WITH POLICYMAKERS TO SHAPE REGULATIONS AND INCENTIVES THAT PROMOTE ELECTROCULTURE ADOPTION.

ACTION STEPS:

- FORM ALLIANCES WITH AGRICULTURAL ADVOCACY GROUPS TO LOBBY FOR SUSTAINABLE FARMING POLICIES.

- SHARE SUCCESS STORIES AND CASE STUDIES THAT HIGHLIGHT THE BENEFITS OF ELECTROCULTURE WITH POLICYMAKERS.

- PARTICIPATE IN PUBLIC CONSULTATIONS AND PROVIDE INPUT ON AGRICULTURAL POLICIES AT LOCAL, NATIONAL, AND INTERNATIONAL LEVELS.

3.4 COMMUNITY BUILDING

PATHWAY: STRENGTHEN THE ELECTROCULTURE COMMUNITY BY FOSTERING COLLABORATION AND KNOWLEDGE SHARING. ESTABLISH REGIONAL AND GLOBAL NETWORKS THAT FACILITATE COMMUNICATION AND SUPPORT AMONG PRACTITIONERS.

ACTION STEPS:

- CREATE ONLINE FORUMS AND SOCIAL MEDIA GROUPS FOR ELECTROCULTURE ENTHUSIASTS TO CONNECT AND SHARE EXPERIENCES.

- ORGANIZE LOCAL ELECTROCULTURE MEETUPS, WORKSHOPS, AND FIELD DAYS TO BUILD A SENSE OF COMMUNITY.

- PROMOTE MENTORSHIP AND KNOWLEDGE EXCHANGE BETWEEN EXPERIENCED PRACTITIONERS AND NEWCOMERS.

DAVID HOLMAN

3.5 SUSTAINABLE FARMING PRACTICES

PATHWAY: IMPLEMENT AND CHAMPION SUSTAINABLE FARMING PRACTICES THAT PRIORITIZE SOIL HEALTH, WATER CONSERVATION, AND ECOSYSTEM RESILIENCE. LEAD BY EXAMPLE AND DEMONSTRATE THE BENEFITS OF ELECTROCULTURE.

ACTION STEPS:

- INTEGRATE ELECTROCULTURE INTO YOUR FARMING OPERATIONS AND MONITOR ITS IMPACT ON CROP YIELDS AND RESOURCE CONSERVATION.

- SHARE YOUR SUSTAINABLE FARMING JOURNEY THROUGH BLOGS, DOCUMENTARIES, AND SOCIAL MEDIA TO INSPIRE OTHERS.

- COLLABORATE WITH LOCAL CONSERVATION ORGANIZATIONS TO PROMOTE

ENVIRONMENTALLY FRIENDLY AGRICULTURAL PRACTICES.

SECTION 4: A VISION REALIZED

4.1 A FLOURISHING EARTH

VISION: PICTURE A WORLD WHERE ELECTROCULTURE HAS BECOME THE NORM, NURTURING THE EARTH'S RESOURCES RATHER THAN DEPLETING THEM. ABUNDANT HARVESTS FLOURISH UNDER THE CARE OF ENVIRONMENTALLY CONSCIOUS FARMERS.

ACTION STEPS: CONTINUE ADVOCATING, INNOVATING, AND EDUCATING. YOUR CONTRIBUTIONS WILL RIPPLE THROUGH GENERATIONS, TRANSFORMING AGRICULTURE AND SECURING A BRIGHTER FUTURE.

4.2 A LEGACY OF SUSTAINABILITY

VISION: ENVISION LEAVING A LEGACY OF SUSTAINABILITY AND RESILIENCE FOR FUTURE GENERATIONS. YOUR COMMITMENT TO ELECTROCULTURE HAS HELPED SAFEGUARD THE PLANET'S RESOURCES AND NOURISHED COMMUNITIES.

ACTION STEPS: REFLECT ON YOUR JOURNEY, CELEBRATE YOUR ACHIEVEMENTS, AND MENTOR THE NEXT GENERATION OF ELECTROCULTURE PRACTITIONERS. YOUR LEGACY IS ONE OF HOPE, PROGRESS, AND STEWARDSHIP.

CLOSING WORDS

AS WE CONCLUDE THIS BOOK ON ADVANCED ELECTROCULTURE, WE INVITE YOU TO EMBARK ON YOUR OWN JOURNEY, ARMED WITH KNOWLEDGE, INSPIRATION, AND A VISION FOR A MORE SUSTAINABLE

AGRICULTURAL FUTURE. ELECTROCULTURE HOLDS THE PROMISE OF REVOLUTIONIZING FARMING PRACTICES, ADDRESSING GLOBAL CHALLENGES, AND LEAVING A POSITIVE IMPACT ON THE WORLD.

MAY YOUR FIELDS FLOURISH, YOUR HARVESTS ABOUND, AND YOUR COMMITMENT TO SUSTAINABILITY ECHO THROUGH THE ANNALS OF AGRICULTURE. TOGETHER, WE CAN CULTIVATE A GREENER, HEALTHIER, AND MORE PROSPEROUS PLANET FOR GENERATIONS TO COME.

THIS CONCLUDING CHAPTER ENCAPSULATES THE KEY TAKEAWAYS FROM OUR EXPLORATION OF ADVANCED ELECTROCULTURE, OFFERS WORDS OF ENCOURAGEMENT, AND OUTLINES A VISION FOR THE PATH

DAVID HOLMAN

TO SUSTAINABLE AGRICULTURE.
IT EMPHASIZES THE ROLE OF
EDUCATION, RESEARCH, POLICY
ADVOCACY, COMMUNITY
BUILDING, AND SUSTAINABLE
FARMING PRACTICES IN
REALIZING THE POTENTIAL OF
ELECTROCULTURE.

THANK YOU

DAVID HOLMAN

www.ingramcontent.com/pod-product-compliance
Lightning Source LLC
Chambersburg PA
CBHW072143290526
45794CB00004B/1409